BUILD YOUR OWN

THE DRILL PRESS

Written & Illustrated

By

DAVID J. GINGERY

Printed in U. S. A.

Copyright © 1982 David J. Gingery
All Rights Reserved

Library Of Congress Catalog
Card Number 80-66142

ISBN 1-878087-04-5

David J. Gingery Publishing L.L.C.
P.O. Box 318
Rogersville, MO 65742

http://www.gingerybooks.com

Email: gingery@gingerybooks.com

CONTENTS

PAGE

INTRODUCTION---------------------------------------   6

CHAPTER

I	CONSTRUCTION METHODS---------------------	8
II	FITTING OUT THE LATHE--------------------	12
III	THE COLUMN AND BASE----------------------	31
IV	THE QUILL FEED MECHANISM-----------------	50
V	THE QUILL--------------------------------	73
VI	THE SPINDLE DRIVE------------------------	89
VII	THE MOTOR DRIVE--------------------------	107
VIII	THE WORK TABLE---------------------------	118

INDEX TO MAJOR SUB HEADINGS----------------------   128

THE METHODS AND MATERIALS THAT ARE SUGGESTED IN THIS MANUAL WERE DEVELOPED BY A NONPROFESSIONAL. THE AUTHOR IS NOT AN ENGINEER OR SCIENTIST. NO LIABILITY IS ASSUMED FOR INJURY TO PERSONS OR PROPERTY THAT MAY RESULT FROM THE USE OF THIS INFORMATION.

FOREWORD

It's true that you can buy a small drill press for a moderate price, but that would deprive you of the fun of building one, and you would not gain the skill and knowledge that will be developed by these exercises.

If you have followed the series in sequence, and completed the lathe, the shaper and the miller, you will have done considerable drilling with your electric hand drill. This is tedious work, and frustrating too, for it takes practice to produce a hole that is in line and on center with a hand drill. No doubt your desire to own a drill press is much greater than when you began.

The patterns and castings for the drill press are not difficult if you have the back ground of the earlier projects, so we'll spend more time discussing machining operations than we will on foundry.

When I began, I expected to make considerable use of both the milling machine and the shaper to machine the castings, but the home made lathe with its 7" swing was easily capable of doing all of the work except milling the keyway in the spindle.

It would appear that the drill press could have been presented earlier in the series because it has such a simple appearance, but there are some exacting requirements in the machining operations that I feel would have discouraged many.

We are still working with simple accessories, and the project is additional proof of the worth of a simple lathe equipped with no more than a face plate and centers. The drill press will be a vital neccessity for building the deluxe accessories though, and so will the additional skill and knowledge that you will acquire wlth this project.

The design exceeds the needs of most home shop operations, for it will drill a 5/8" hole in steel with ease. It can be scaled down to a smaller size easily, or you can scale it up to a hefty floor model if you want to. Keep in mind that some of the parts use the extreme capacity of the lathe though, so you will have to depend on the larger swing of the miller or a larger lathe if you scale up.

The drill press is likely to be the greatest challenge yet. It will certainly be the most used machine in the shop, and you will be a far better machinist and mechanic when you have built it.

Slate's Sensitive Drill.

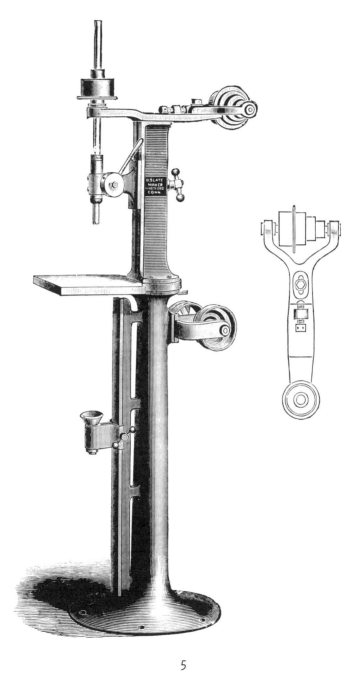

INTRODUCTION

    The engravings of old machines in earlier manuals have been an aid in describing the design and function of each machine. Modern machines have been much improved, but the essential elements of design remain the same. I have some fine old engravings of " camel back " drilling machines, but I think Slates " Sensitive " drill press is the best illustration for discussion in this project. It has interesting features that are obvious, but there are design considerations that might escape notice if not pointed out. Like the other engravings, this one is from an 1888 Hill and Clarke catalog.
    Note that the bracket which supports the driven pulley and the cone pulley is cast and machined separately. It was possible to adjust the bracket to bring the spindle exactly perpendicular to the table and parallel to the column. This is the first vital design consideration in building an accurate drill press.
    The table could not be raised or lowered, but the quill support would be moved up or down for the same purpose.
    The spindle was mounted in close fitting bearings in the quill, and its upper end passed through the hub of the driven pulley which was mounted in its own bearings. No belt tension was transferred to the spindle. It is most likely that the bearings were poured babbit, though it may have had a ball thrust bearing on the quill.
    The feed lever rotates a small pinion gear which fits a rack milled in the quill, and the quill support is split so that it can be adjusted closely to the quill after it wears in use.
    The table could be swung aside so that a shaft could be set in the cup center for accurate center drilling. An additonal accessory clamp was available for gripping the upper end of the shaft.
    The machine sold for $80.00 in 1888, and it was offered as a bench model without the pedestal for $60.00. That was a pretty high price for such a simple machine in those days, and the reason is the amount of hand work that was neccessary to finish it to precise standards. I think we have advanced more in the design of drilling machines than in any other of the machine groups. You can buy a small drill press complete with motor for about $150.00, and it is a better machine than this early model. Such a

comparison in other machine groups shows that there must have been a lag in interest and lack of competition in the manufacture and marketing of drilling machines, for lathes and millers and shapers were remarkably cheap by comparison. It was not until the round column appeared that it became possible to mass produce drilling machines cheaply.

Slate's machine was termed a " sensitive " drill because the operator could feel the amount of feed pressure with his hand. It is at the moment that the drill point cuts through the material that small drills break, and a reduction of feed pressure is needed at that moment. If there is excess vertical play in the spindle the advantage of a sensitive hand feed is lost.

Our drill press will look much different, but the same considerations must be kept in mind as you make each part and assemble them. As the bed is the foundation of the lathe, the column is the foundation of the drill press. All of the members of the drill press are either parallel to the column or at exact right angles to it. With care, you can produce the parts to near perfection, and you can easily install shims to correct for small errors if you have not yet developed the keen eye and sensitive fingers. You will certainly grow in skill as you machine each part, and your simple lathe will win your respect as it performs these operations without the aid of expensive accessories and tooling.

The machining operations are turning, boring and facing off, and you will have acquired these skills in earlier projects in the series. The important variations are in setting up the work and holding it for machining. You will gain some new simple accessories for future projects as you fit out to machine the parts for the drill press.

All of the patterns are of simple shape and easy to mold and cast. They are all within the capacity of the charcoal foundry using the one quart pot.

Like the other home made machines, the project may be a bit intimidating, but I urge you to go ahead and try out your skill. Once you have the base and column built you are not likely to want to stop. It may take six weeks, or it may take six months. In either case you will not have a series of monthly back breaking payments to make, and you will own still another fine machine you have built.

CHAPTER I

CONSTRUCTION METHODS

It will be the boast of every manufacturer of a modern drill press that the head of his machine is cast in one piece and that it is precisely machined so that the column and spindle are exactly parallel. He is justly entitled to his boast, for the simplicity, durability and accuracy of modern machines is amazing. The product is a tribute to the skill and genius of the engineer, pattern maker, molder and machinist.

It requires great skill, knowledge and experience to design a pattern for the mass production of such an item as the head for a drill press. There are hundreds of details to consider, and a single error can create a giant heap of scrap iron. Our job is much easier because we are only building one drill press, and we're not going to try to cast the head in one piece.

FOUNDRY WORK

Having separated the major parts of the machine into manageable sub assemblies, there are no difficult patterns to make and most of the molding is routine practice. We won't have to discuss it in detail except to review some of the basic principles when the need arises.

PATTERN MAKING

All of the patterns are simple rectangular or cylindrical shapes, or combinations joined with glue and brads. It's worth the time and trouble to produce accurate and smoothly finished patterns because the wood of the pattern is much easier to work than the metal of the casting.

Draft is the all important factor, but it must not be over done. Very shallow patterns can be made with little or no draft because rapping will enlarge the cavity enough so that the pattern can be cleanly drawn from the mold. A pattern with depth of 3" does not need a greater angle of draft than one of 1" depth, but the draft is much more important on the deeper pattern. With few exceptions, draft of one degree is more than adequate for the patterns in this project.

The nominal dimensions in the pattern drawings refer to the size of the pattern at the parting plane. It will usually be best to cut the pattern to the nominal size, and then form the draft with a sanding block. All vertical surfaces slope slightly to a smaller dimension than at the parting plane. Test the angle of draft by holding a small try square against the parting plane to make sure the vertical surfaces slope slightly.

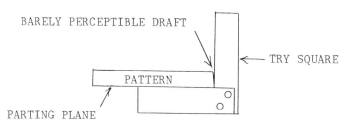

When more than one vertical rib or webb forms a channel, their axis must be at exact right angles to the parting plane so the pattern won't be locked in the mold.

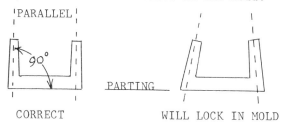

Fillets are the slightly rounded junction at inside corners. They leave a cleaner cavity and relieve strain at the corners when the metal solidifies, but if they are made too large they cause shrink cavities. Outside corners are slightly rounded except at the parting plane.

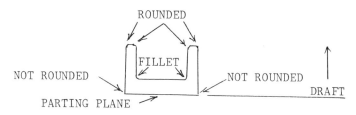

The direction of draft will be indicated by an arrow above the word " DRAFT ". Outside vertical surfaces slope inward, and inside vertical surfaces slope outward.

Fill any flaws with plastic body putty, and sand all surfaces smooth. Seal all patterns with two or more coats of lacquer, varnish or shellac. I use clear lacquer in an aerosol can because it dries fast.

A ROUTINE MOLD

These simple steps go into every mold, and we'll discuss variations when the need arises.

Lay the pattern on the molding board with its parting plane flat on the board.

Dust both the pattern and the board with parting compound.

Invert the drag half of the flask over the pattern, and ram up the drag.

Strike the drag off level, vent with the wire, rub in a bottom board and roll over the drag.

Set the molding board aside and clean up the parting face of the drag.

If it is a split pattern mold, set the cope half of the pattern in place, press in the sprue pin, set the cope on the drag, ram up the cope and strike off level.

If it is a simple pattern, just set the sprue pin and the cope, ram up the cope and strike off level.

Remove the sprue pin and finish the sprue opening to a funnel shape.

Vent the cope with the wire and lift it straight up and set it on edge behind the drag for a simple pattern.

If it is a split pattern, rub in a bottom board on the cope so you can lay it down to rap and draw the pattern.

Swab all around the pattern, drive in a draw peg, rap the peg in all directions, and lift the pattern straight up in a single motion.

Cut the gates and runners, clean up the cavity and swab the gates and runners and any weak areas.

Blow out the sprue with the bellows, swab the bottom of the sprue opening and check for possible errors.

Turn the cope horizontal before you move it over the drag to close up the mold.

Close up firmly, but don't drop the cope or jar the mold.

Pour directly into the sprue opening.

COMMON PROBLEMS

Aluminum castings require sprues and risers about three times as large as iron castings. You could hardly use a sprue that was too large unless you exceed your melting capacity, but one that is too small will cause a shrink flaw in the casting.

Too much water in the sand and failure to vent with the wire will cause depressions in the surface of the casting that look much like shrink cavities. Ramming too hard can produce the same effect.

If the sand is too dry, or if you allow the mold to stand too long before pouring, you may carry loose sand into the cavity when you pour.

Newly tempered sand is difficult to work because the moisture isn't uniformly spread throughout the sand. It improves a great deal after standing for several hours. I always shake out my castings directly into the riddle placed in my sand bin. Then I riddle the sand to leave only the hardened lumps in the riddle. These I break up and riddle into the bin, sprinkle to replace lost water, and mix it up with the main body of sand. When I close up shop I cover the sand bin wlth plastic to prevent evaporation so my sand is ready when ever I need it.

Fill the flask in progressive layers of about 2" thickness, and ram each layer uniformly. If there are soft spots or empty corners it's because you were not systematic and thorough with your ramming. Press sand into the corners and small core areas with your fingers before you begin general ramming.

If you attempt to slick up the parting face of the drag after the cope is rammed you are likely to have a run out at the parting line when you pour. Don't disturb the face of the drag once the cope is rammed up.

Pouring to the side of the sprue will wash loose sand into the cavity. Pour directly into the opening as fast as it will accept the metal, and don't stop until it is completely filled.

If gates and runners are too small it has the same effect as an undersize sprue or riser. They should be from 2/3 to 3/4 the volume of the sprue or riser, and as short as practical.

Use a light to examine the cavity, and swab any areas that threaten trouble. Dust newly swabbed areas with parting to harden them up.

CHAPTER II

FITTING OUT THE LATHE

These simple accessories will handle all of the work that would normally require a three or four jaw chuck, and they will be a valuable addition to your shop for future projects.

A PLAIN ANGLE PLATE

Some of the castings can be bolted directly to the face plate to machine them, but others will have to be mounted at right angles to the turning axis of the lathe. For this you need an angle plate.

Only one pattern is required, but it takes an angle plate to machine another one so you'll have to cast two of them from the same pattern.

The shorter leg of the angle is the mounting leg, and it needs to be rounded to clear the bed when mounted near the rim of the face plate. I realized this a bit late, so you'll notice that those in the photos have not yet been rounded off. I had to do it the hard way later.

THE ANGLE PLATE PATTERN

Two simple shapes of white pine joined with glue and brads. Only a minimum of draft is required, and wipe a very small fillet at the inside corner of the angle.

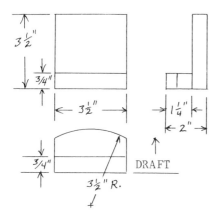
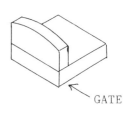

12

MOLDING THE ANGLE PLATE

This is a routine mold with the entire pattern in the drag. Ram up the pattern in the drag, roll over and set a 1 1/4" sprue pin about 1" away from the base. Ram up the cope, open the mold to draw the pattern and cut the gate, and close up to pour.
Make two molds from the same pattern and pour them at the same time. It will be more economical to operate your foundry if you prepare a number of molds to be poured with each melt so that you use up the entire fuel charge.
As the fuel bed settles, add charcoal in single layers frequently to maintain the height of the bed. To add large amounts will cool off the fire and waste fuel.
The broad flat surface of this casting will be slightly concave because of normal shrinkage. A machining allowance has been made so there will be ample thickness after it is faced off smooth.

MACHINING THE ANGLE PLATES

The final accuracy of the angle plates will be dependent upon the trueness of your face plate and the lathe cross slide.
Rotate the face plate to see if it wobbles. If it does it is bent and you must straighten it and face it off to true it up.
Test the surface of the plate with a good straight edge to see if it is concave or convex. If your cross slide is traveling at exact right angles to the turning axis it will be perfectly flat when faced off. Correct any error by scraping the front vertical pad of the carriage box slide.
With the lathe at normal standards of accuracy you are ready to begin machining the angle plates.
Drill and tap four 5/16"-18 holes in each casting as shown in the drawing.

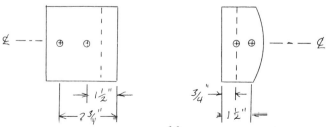

These tapped holes will serve for mounting the castings for machining, and will also be used in the different operations on the drill press castings. Other holes can be drilled or tapped for future jobs.

The first step is to face off the large surface of the castings. Just bolt the casting to the face plate with a cap screw through the face plate, using a bushing or some set screw collars in a stack for spacers. One bolt will clamp it securely enough to take moderate cuts as you make several passes to bring the surface smooth and flat.

No attempt is made at this point to make the base angle true, and the finished thickness is not critical. Even if the thickness finishes up at as little as 1/2" the plate will be strong enough. Face off both castings, and there is no need to make them of equal thickness.

The second step is to mount the base of one of these faced off angles to the face plate with the machined surface at exact right angles to the face plate.

Because of the draft the angle will be greater than 90 degrees, so you need to add sheet metal shims to bring it true square with the face plate. File the corner of the angle so it rests firmly on the face plate, and use a single bolt through the face plate slot into the tapped hole in the base that is nearest to the center. Test with an accurate square, and add shims until the angle is true.

Re-check the angle after tightening the bolt firmly, and be careful not to deform the face plate by over-tightening the mounting bolt.

When the angle is true, bolt the second casting to it with two 1/4" bolts and face off the mounting base.

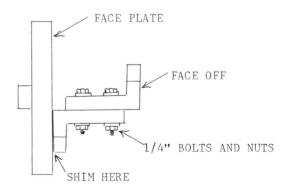

This will give you one true angle plate, and you need only bolt the faced off base on the face plate without the shims to machine the second angle plate true square.

If there has been an error of more than a tiny fraction of a degree it will be apparent when you rest both bases on a true flat surface. A very small error can be corrected by hand scraping, but the job should be re-done to correct a serious error.

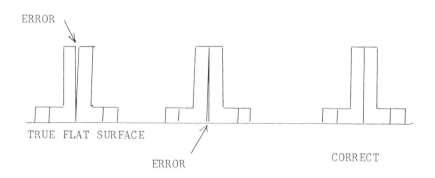

ERROR

TRUE FLAT SURFACE

ERROR

CORRECT

The angle plates will be more valuable if all surfaces are made true square and parallel. You can use either one to mount the other for machining the edges.

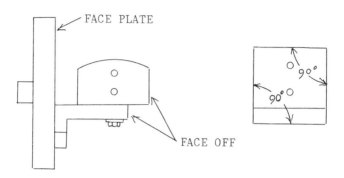

The quill guide is one of the casting assemblies that is mounted on the angle plate for boring. Note the weights that are added for counterbalancing the work.

A " V " BLOCK ANGLE PLATE

   A number of smaller parts need to be bored accurately so that they can be mounted on an arbor between centers. The sprue forms a handy shank for mounting them in the V block angle plate. A starting hole is drilled, they are bored to size, a set screw is installed and they are cut from the sprue to be mounted on the arbor.

   The main body pattern is split for easy molding, but the clamp bar is a simple one piece pattern.
   The inverted V on the clamp bar serves as a truss to stiffen the bar for large work, and provides a better grip on small work when it is turned over.
   Make the patterns of 1/2" thick pine stock rather than plywood for easier working and finishing.
   The drag half of the pattern differs from the plain angle plate only in dimensions. Make it as near squared at the corners as possible to simplify aligning the cope half parts.
   The V blocks that form the cope half should be made as near identical as possible so that work of different diameters will be gripped in alignment with the turning axis of the lathe.

THE DRAG HALF PATTERN

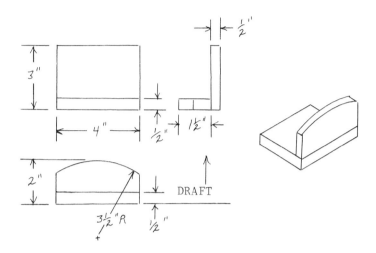

THE COPE HALF PATTERN

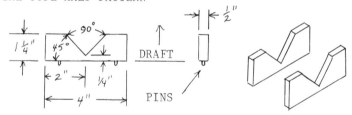

Two small pins made from brads are pressed into the base of cope parts to register with holes in the drag half.

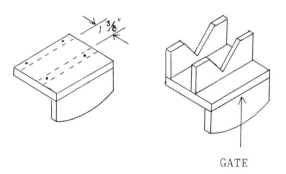

GATE

MOLDING THE SPLIT PATTERN

Ram up the drag half pattern just as you did for the plain angle plates. Vent with the wire, rub in a bottom board and roll over.

The pins in the cope half are meant to fit freely so that the parts will separate when the mold is opened.

Set the cope parts in place, set a 1 1/4" sprue pin about 1" away from the base and ram up the cope.

Remove the sprue pin and cut the sprue opening to a funnel shape. Vent the cope, rub in a bottom board and open the mold.

Lay the cope down on the bottom board to swab, rap and draw the pattern parts. You should be able to grasp the pins with a small pliers to lift out the pattern parts.

Drive a peg into the drag half pattern, swab all around the pattern, rap and draw it from the mold.

Cut the gate, clean up and finish carefully, and close up to pour.

THE CLAMP BAR

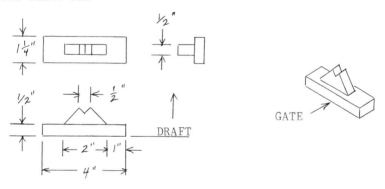

The inverted V block has an included angle of 90 degrees, and so does the small V.

Molding is routine as for any simple pattern, and the mold is fed with a 1 1/4" sprue.

MACHINING THE BASE

A 1" diameter by 8" long arbor will be needed for some of the machining operations, and it will serve to hold the V block angle plate to face off the mounting base.

Drill a 1/4" hole centered 1/2" from each end of the clamp bar, and use the bar as a guide to drill matching 1/4" holes in the main body of the angle plate.

Tap the holes in the angle plate for 5/16"-18 threads, and enlarge the holes in the clamp bar to 3/8".

File the V notches in the angle plate to bring them as near identical as possible. You can use a large piece of square stock as a guage.

Clamp the fixture to the arbor and mount it between centers on the lathe to face off the mounting leg.

Drill and tap two 5/16"-18 holes in the base on the same dimensions as for the plain angle plate.

When bolted to the face plate, work clamped in the V blocks will be held parallel to the turning axis of the lathe.

In order to be a truly precision fixture the V notches would have to be absolutely identical, but it will align work within a fraction of one degree if you use reasonable care as you fit them by hand. This will be adequate for the needs of this project and most operations that are to be done in the home shop.

DRILLING, TAPPING AND BORING WITH THE LATHE

Many of the holes can be drilled and tapped by hand, but a few must be done on the face plate using the tail stock chuck. You can order a 1/2" chuck with a #1 Morse taper adapter from Wholesale Tool Co., and it will provide the needed accessory for the tail stock and also the chuck for the drill press when it is complete.

In the photo below, the work table clamp base is bolted to the face plate with a short length of pipe as the spacer so that the faced off surface will be parallel to the column.

A small dimple was cut in the center of the faced off surface to start the drill, and a 5/16" tap size drill is held in the tail stock chuck and fed into the work with the tail stock screw. The vise grip pliers prevent the shank of the adapter from turning.

The same method is used to drill the starting hole for holes that must be bored to a precise size.

A bit tedious as compared to drilling with a drill press, but it is the only way we have until the drill press is complete.

Taps can be held in a tap wrench or in the tail stock chuck. The face plate is rotated by hand for tapping.

All operations are completed without changing the original set up on the face plate, so the faced off surface will be parallel to the column, and the tapped hole will be exactly perpendicular to the faced off surface.

Work that is to be bored is mounted either directly to the face plate, with spacers to give clearance for holes that are bored through, or with on of the angle fixtures.

There will be more time spent in boring than in any other phase of construction. These operations will determine the ultimate accuracy of the drill press, so they must be done with great care and patience.

BORING TOOLS

The tool post of the home made lathe will hold a 1/4" forged boring tool in its slot. When extended as much as 2 1/2" to 3", the tool is very flexible and there is much tendency for it to spring away from the work. The support at the tool post and the form and condition of the cutting edge will have a great effect. Much time can be saved by grinding to proper form and honing to a razor sharp edge.

The photo shows the column base casting being bored with the tool mounted in the tool post with only washers for support. When I added the simple V block to support the tool I was able to make deeper cuts with each pass and the cut was much smoother because chatter was eliminated.

You can buy forged boring tools but they are quite expensive at $6.00 to $7.00 each. A 36" length of 1/4" drill rod will make a half dozen or more for less than the price of one.

There is little point in making a 1/4" boring tool any longer than 5" or 6" because they are not rigid enough for deep hole boring.

It's a simple matter to heat the end of the rod to a bright red heat and forge a short hook. You heat it to a bright red again and quench it in oil or water to harden it. Then you polish the hardened end with emery and heat it until it turns a light straw color and quench it to temper it.

Commercially made boring tools are hardened and tempered for the entire length, and that makes a more rigid tool, but it is difficult to do without a muffle furnace to heat the entire tool evenly. I harden and temper only the working end on my home made boring tools.

The conventional boring tool is bent to a short hook and ground to a form much like a turning tool.

Such a tool works very well, but it won't enter the small starting hole to finish a bore at less than about 3/4". A smaller tool is made by heating the end to bright red and swelling it with a few blows of the hammer.

After hardening and tempering, it is ground to a " D " shape and given the proper angle and cutting edge. Such a tool will enter a small starting hole, and it will work very well on large bores too.

BORING TECHNIQUE

Learning to bore precisely requires study and patience. The slightest error in tool form or set up can make a great deal of difference. Mysterious things may happen in your early experience, but they will be found to have a simple explanation.

Given the proper form and set up, the tool will cut well only if it has a very keen edge. That requires wet honing to razor sharpness in addition to grinding to form.

The cutting edge must be on the center line of the bore for accurately predictable cuts. If it is above or below center the cut will be deeper than predicted.

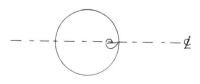

The cutting edge needs a slight amount of " rake " to cut properly. That means the top surface of the tool is slanted. If the rake is too slight the cutting action is poor and the edge is quickly blunted. If the rake is too great the tool will chatter and produce a rough cut.

2 TO 4 DEGREES OF RAKE

RELIEF

The cutting edge needs " relief " for clearance below the cutting edge. If there is not enough relief the effect is similar to not enough top rake: Poor cutting action, if any, and the tool is quickly blunted. Too much relief causes chatter and a rough cut.

It can take several evenings of study and practice to learn how to grind and sharpen tools. It will be worth your while to provide a couple of extra castings for practice so that you can master this simple skill.

PREDICTING THE FINISHED SIZE OF THE BORE

Each of the required bores will need to be finished to an exact size so mating parts will fit without play. Since the home made lathe lacks graduated collars on the feed screws, this may seem to be an impossible chore. Actually, graduated collars would be of little value if you had them. The graduations only provide an accurate estimate of the depth of cut, and they are not able to take into consideration the yield of the tool and the work. You can estimate just as well without graduations.

Actually, there are clear graduations even though you have no markings on the collar. Since one full turn of the feed screw will advance the tool a predictable amount, then a fraction of a turn is also predictable. It is easy to note the position of the feed handle and rotate it just one half turn, and just as easy to rotate it just one quarter. This gives you four reasonably accurate divisions if you know how much the tool is advanced with a full turn of the screw. It does not take long to learn to divide the quarter turns into yet smaller increments so you can predict a cut of as little as .002" with amazing accuracy.

We've discussed it in earlier manuals, but it can be an elusive process of thought until it is well understood. We'll review it here in case you still need a little help.

You need only know the pitch of the feed screw and you know how much the tool is advanced with one turn of the screw. It requires 20 turns of the screw to advance the tool one full inch on the home made lathe. That means it is a 20 pitch screw, and one full turn moves the tool one twentieth of an inch. Other lathes may have 8, 10, or 16 pitch screws, and you can easily find out by counting the number of turns that move the tool one inch. Then apply the fraction to determine the amount of feed for one turn.

Convert the fraction to a decimal number to determine how many thousandths of an inch each turn represents. On the home made lathe 1/20" is equal to .050". (One divided by twenty.) One half turn represents .025", one fourth turn represents .0125", and one eighth turn represents .00625".

Keep in mind that these figures represent the travel of the tool, not the depth of cut. As you are moving the tool on the radius of the cutting circle, the effect will be twice as great on the diameter. On the home made lathe one fourth of a turn will advance the cutter .0125" and it will change the diameter by .025". For lathe work, it is neccessary to reason in terms of depth of cut rather than the amount of tool travel.

Whether or not the screw has graduations on the collar, the predicted depth of cut is purely theoretical. It does not take into consideration the yield of the tool or the work piece, and the measured result will not be what was indicated in most cases. In actual practice you will make a series of cuts to clean up the work. Then you advance the cutter to make a theoretical cut of .025". You measure the work after the cut to find that you changed the diameter by only .023". This means that the tool or the work piece, or both, yielded by .002". An additional pass at the same setting may bring the work to size, or you may have to advance the cutter an additional amount to compensate for yield on the second pass.

The difference between the theoretical cut and the actual cut is much greater with the small forged boring tool because it lacks firm support near the cutting edge. You measure the bore before and after each pass to learn the exact amount that will be cut at any theoretic setting of the cross feed.

Let's discuss a practical job to demonstrate the principles involved. A 1" X 8" arbor will be needed for machining some of the parts for the drill press. It must be a truly round arbor that runs concentric with the axis of the lathe, and its diameter should be no more than 1" and no smaller than .001" undersize. The length is not critical.

You could not possibly locate and drill the turning centers precisely on a 1" shaft, so you must begin with a larger shaft and turn it to dimension between centers on the lathe.

An 8" length of 1 1/8" or 1 1/4" cold rolled steel is prepared with 60 degree centers in each end, and it's mounted between centers with a dog to drive it.

It will not be perfectly centered, so one or more passes will have to be made to bring the shaft concentric. Nothing about the actual cutting depth can be determined from these " cleaning up " passes, but you can measure the cleaned up diameter at both ends to see if your centers are properly aligned. If one end is larger than the other, you must adjust the tail stock set over to bring the turning axis parallel to the tool travel.

You can only machine from 2/3 to 3/4 of the length of the shaft because the dog takes up space on the end. The work will be turned end for end and remounted to machine the remaining length when the first portion is finished.

You will need a micrometer or a vernier calipers for precise measuring. My own preference is the vernier calipers because it measures both inside and outside, and a single tool covers a wide range of sizes. You can get a nice stainless steel one form Wholesale Tool Co. for less than $20.00.

When the shaft is cleaned up, and of uniform diameter from end to end, you can measure it to establish the starting dimension. Let's say you began with a 1 1/4" diameter, which is 1.250", and it cleaned up at 1.205". You must reduce it by an additional .205" to finish at 1.00".

The position of the feed screw handle now becones the reference, and it is easiest to judge it when it is either in a vertical or horizontal position. Then it is easy to turn it one quarter turn from horizontal to bring it vertical, or one quarter turn from vertical to bring it horizontal. The object now is to make a pass at a setting that will finish at some even multiple of .025" with the feed handle in either the vertical or horizontal position.

Let's say that the handle was exactly vertical when

you made the last cleaning up pass. The cleaned up dimension is 1.205", and if you advanced the tool 1/4 turn to remove an additional .025", the theoretic dimension after the pass will be 1.180". If you try to estimate a fractional turn of the handle to finish at 1.175" it will not be exactly horizontal, and you won't have an accurate reference point. This is where the compound feed screw gets into the act.

The compound feed screw on the mome made lathe has the same pitch as the cross feed screw, but because it is set at an angle to the work it doesn't change the depth of cut as much per turn.

Turn the cross feed screw 1/4 turn to horizontal, and turn the compound screw about 1/16 turn and make a trial pass. Measure the result to see how close you came to the theoretic 1.175".

Always turn off the lathe and wait for the work to stop before you make any measurement.

If you finished the trial pass at something like 1.177" it is only neccessary to advance the compound a very tiny amount to come to an even dimension. If you finished at near 1.173" the compound must be backed up.

There is always a small amount of end play in a feed screw, and when you need to back it up you must move it enough to take up the end play and an additional amount to back up the tool.

Maybe the next pass will finish up at 1.151", so just advance the cross feed 1/4 turn and advance the compound a very tiny amount to attempt to finish the next pass at exactly 1.125". You still have at least four more chances to get it right before the final pass.

By these trials you will learn to judge the depth of cut very accurately, and you will be making adjustments in " teeny bits ", " tads ", " smidgeons ", and such terms that might seem a disgrace to some, but you can work to a tolerance of plus or minus .001" without graduated collars, and they can't.

The same method applies to boring, but multiple passes are made on each setting. You must measure after each pass and learn the effect of each setting. It can take as many as five or six passes to make a theoretic cut of .025" in aluminum, and even more in steel.

Let's say you nave cleaned up the bore of the column base casting at 1.525", and you want to enlarge it to the size of the column, which is 1.650". That's .125" to be removed, so you have plenty of room to work with before the final pass.

Advance the screw 1/4 turn for a cut of .025" and make one pass to the bottom of the bore and withdraw the carriage. Measure the bore carefully, and you will likely find that you have only removed about .012" because the slender tool springs away from the work.

Make a second pass without changing the setting of the screw, and you will probably have removed an additional .006".

A third pass may remove .004", a fourth pass .002", and a fifth pass .001".

A series of trial passes with careful measurement after each pass will teach you what to expect, and this must be learned with each new set up because the tool will not perform in the same way when the set up is changed.

Once you have established the setting and determined the number of passes for a given set up there is no need to measure after any but the last pass of each setting.

The only precision tool I've used in building my own shop is a $13.00 vernier calipers that enables me to measure to .001". I haven't put graduations on any of my feed screw collars because I simply don't need them. I get sadistic pleasure out of machining a piece to within .001", and then handing it to the bird who says it simply can't be done.

DON'T GIVE UP

Many of the operations in building these projects are tedious and time consuming, and it's easy to get discouraged. Speed and skill come with practice though, and you simply can't acquire them in any easy way.

I had to mold my lathe bed six times before I got two good castings, and I'm not a patient man. I would have stopped then if it were not for the commitment to write this series of manuals, because I really didn't want a lathe as much as I thought I did when I began. All of my machines are finished now, and I have a project list ahead with no end in sight. I'm glad I didn't quit.

Many of my future projects are items I would have felt beyond me when I began this series. Having the machines helps make them possible, but, more important, I've learned how to solve problems. That really is the most valuable part of the whole experience.

CHAPTER III

THE COLUMN AND BASE

The main feature of a drill press that makes it so valuable is its ability to drill a truly perpendicular hole exactly where you want it. Extra care must be taken to be sure that the base and table are at right angles to the column, and that the quill and spindle are parallel to the column.

DRESSING THE COLUMN

The column of a commercially built drill press will be accurately machined and ground to a high finish, and this job would be beyond the range of most home shops. By using V block clamps instead of the conventional bored clamps we can tolerate a column that is not true round and uniform in diameter over its entire length. Ordinary water pipe will do the job, and it is only neccessary to file the surface and polish it with emery cloth or sand paper.

I've used standard weight pipe for my drill press, but you can get extra heavy pipe for a stronger column. It is the same outside diameter, but the wall thickness is about 1/3 greater. I note a slight deflection of the column on my drill press when drilling large holes in steel, though it would only be a serious problem in very high class work. You should consider extra heavy pipe if you plan on much heavy duty work. Black pipe is much easier to polish than galvanized.

As detailed in this manual, the column is 1 1/4" pipe. You can use larger pipe for your column, but then you must enlarge the column base and re-design the clamps for the head and work table. The clamps are designed to contact the column at opposing points, and the V should be broadened for larger pipe.

There is no reason why you can't use a tall column and a larger base to build a floor model. If you do, I suggest a 2" extra heavy pipe column and a larger base. You can use 1/4" X 2" angle to frame the base, widen it to 12 3/4" and use three slabs of 1/4" X 3" cold rolled steel for the top surface with two slots, and make it 18" deep. Move the column base about 4" forward of the rear edge of the base for better balance, and bolt it to the floor. Add a brace made of angle iron bolted to the underside to span the top irons.

Dressing the column is a simple job if you have a lathe with enough capacity between centers. You would just mount it with a pair of pipe centers, draw file it and polish up with emery cloth or sand paper.

In the past I've done the job on a wood lathe by preparing a pair of hardwood centers which I drove into the ends of the pipe to mount it on wood turning centers. It is extremely dangerous and should not be done. If one of the wooden centers breaks or slips off, the pipe becomes a deadly missile. When spinning work leaves the lathe centers, it will either miss you and scare you near to death, or it will hit you and kill you. It is not likely to be a minor injury at best. Having been fortunate enough to only be scared half to death, I devised a safer method.

It's a simple arbor mounted on a pair of pillow blocks, and you are likely to have most of what you need on hand. The shaft and the motor at least will be used to build the drill press, and the remainder amounts to little cost.

The base is 3/4" plywood, and the supports for the pillow blocks are 6" lengths of 2" X 4" lumber. The supports are glued and nailed to the base, and the pillow blocks are mounted with 1/4" carriage bolts that pass through the base and the supports. Lag bolts would not be safe because the wooden supports could break away.

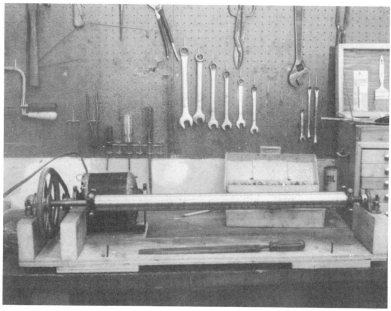

A 12" X 18" piece of 3/4" plywood was added for the motor mount, and another small piece was added to the opposite end so it would rest evenly on the bench.

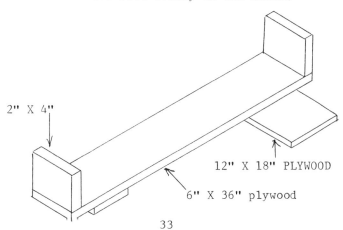

I purchased a 36" length of 5/8" cold rolled shaft, half of which became the spindle for the drill press. A Pair of 5/8" set screw collars take up end play.

I borrowed the 7" pulley from the shapers countershaft, and used a 2" pulley on the 1725 RPM motor to give a working speed of less than 500 RPM.

The 1 1/4" pipe is prepared for mounting by drilling four 1/4" holes 3/8" from each end and tapping them 5/16"-18 for the adjusting screws. This provides a means to adjust the pipe to near center on the shaft.

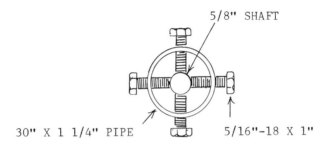

30" X 1 1/4" PIPE   5/8" SHAFT   5/16"-18 X 1"

It does not have to be perfectly centered, and it's not difficult to bring it very close to true.

There is danger from the spinning screws because they become nearly invisible in motion. They can take off a knuckle or a whole finger in a flash. Don't go near such a monster wearing a neck tie, long sleeves, or loose fitting clothing of any kind.

26" to 28" makes a practical height for a bench drill press. One inch on each end will be wasted when you cut away the tapped holes.

There is a protective coating on black pipe that will clog up your file, so dress it up with coarse sand paper before you begin to file. Use a large single cut file of at least 12" length. 14" or 16" is even better. Don't allow the file to ride the work, but move it as though it were held stationary in the vise. Lift it on the back stroke so you don't wear out the teeth. Make a series of diagonal cuts along the entire length to keep the diameter uniform. It's not possible to bring it true round, and the only object is to remove the scaly surface and drastic high areas to improve its surface a bit. The whole operation only takes 10 or 15 minutes, and you can finish by polishing with coarse emery cloth or sand paper.

When it is dressed and polished you will have removed only .002" or .003" of material from the high areas.

You can cut partly through the spinning pipe with a hack saw about 1" from each end. Be very careful of the spinning bolts, and don't cut more than about 3/4 through. Finish cutting off the ends with the column held in the vise.

THE COLUMN BASE CASTING

This pattern will be easy to make if you drill a pair of diagonally opposite holes in the flange and mount it on the lathe face plate to bore the tapered hole and finish the circular part. When it's finished just fill the holes with body putty.

It requires only minimum draft on the outside surfaces, but the bore tapers from 1" diameter at the base to 1 1/2" at the top so it will leave a clean green sand core in the mold.

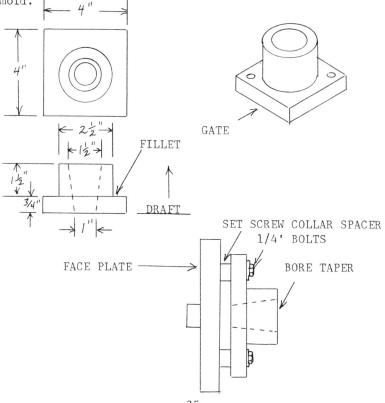

This is a routine mold, fed with a 1 1/4" sprue at any side. Press the core area full of sand with your fingers as you fill the drag, but don't ram it any tighter than the main body of the drag. If the core is rammed too tight it will not yield when you rap the pattern, but will break away and lift out when you draw the pattern. It must be resilient enough to yield when you rap, but firm enough to retain its shape when the pattern is drawn. Take extra care to clean up the cavity to avoid any sand inclusions in the casting. Swab the gate, both top and bottom of the sprue opening, and any weak areas at the edge of the core.

MACHINING THE COLUMN BASE CASTING

If you are not skilled at boring on the lathe, you will be when you have finished this casting. Don't be upset if you have to spend two or three evenings at it. Forget, for now, that you are building a drill press. Just use as much time as you need to learn how to machine this one part. Don't work late until you are bleary eyed and tense. It's not a bad idea to make two castings, or even three, so you will have another on hand in case you spoil one by boring oversize. Nothing will be lost but a little time, and you will gain a valuable skill.

The column will not be perfectly round, so you must determine its average diameter and bore to .001" over the average size. Carefully measure the diameter of the column at several places around it to determine the largest and smallest dimension. Mine measures 1.662" at the smallest and 1.667" at the largest, so it is .005" out of round. I expected worse, so I was pleased. I bored the base to 1.665", which is just one half thousandth over the average, and it was just a bit too snug when I drove it onto the column. I needed no set screws to lock the column, but you could add one or more set screws if you bore just a bit too loose.

You could mount the casting with spacers and bore it all the way through, but you will be more certain that the column is exactly perpendicular to the base if you mount directly to the face plate and bore to a 1/8" shoulder at the bottom of the hole.

Drill four 1/4" holes in the flange, but tap just two of them diagonally opposite 5/16"-18. The tapped holes will be used to mount it on the face plate, and all four will be used as a guide to drill the tap holes in the base later. Then the base is tapped 5/16"-18, and the holes in the casting are enlarged to 5/16" for the bolts.

The first operation is to face off the bottom, so bolt it to the face plate with set screw collars or bushings of equal length for spacers.

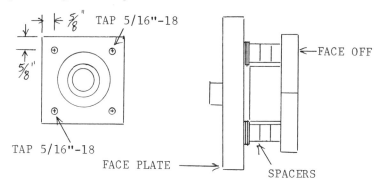

The object here is to machine the bottom true flat so it will mount firmly for the boring operation. You can use a stack of set screw collars and flat washers for spacers, or a pair of bushings or sleeves of equal length, so that the bottom surface will be parallel to the face plate. Center it on the face plate reasonably close, face off the bottom surface, and enlarge the bore to about 1 3/8". A 1/8" shoulder will be left at the bottom of the bore, so there is no need to be precise at this point.

Test the faced off surface with a good straight edge to be sure it is true flat, then mount the machined surface on the face plate without spacers. Center the outside diameter of the circular portion well so that the wall thickness of the boss will be uniform after boring.

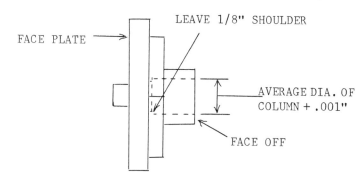

Face off the end of the circular portion, then proceed to clean up the bore to within 1/8" of the bottom. You can make one or more groups of passes at a cutting depth of .050" until the bore is cleaned up, but reduce the depth to .025" when you are within .100" of finished size. Use notes to keep track of what you are doing, and measure as many times as you need to to get consistent readings with your calipers. When you can get the same reading two or three times in a row you are doing it right. Ultimate success depends almost entirely on your ability to measure accurately.

You can make a simple V block to support and stiffen the boring tool from 1/8" X 1/4" key stock. Cut a shallow groove in each of two blocks with the hack saw. Clamp them together in the vise and step drill through the center to about 7/32". Then file the half hole to a V shape. I had to file a shallow notch in the underside of mine to clear the lower edge of the slot in my tool post.

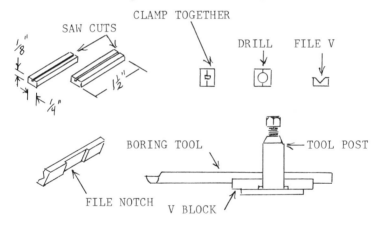

Usual practice is to clamp the tool with two blocks, one on the bottom and the other on top. The slot in my tool post is not deep enough for both blocks, but the one block supporting the lower surface of the bar stiffens it greatly.

Measure the height of the tail center and set the cutting edge of the tool at the same height. Extend the tool just slightly more than enough to reach the bottom of the bore. Hone the tool to razor sharpness, with a tiny radius at the nose.

The intention is to bore the casting to a forced fit on the column. The four holes in the flange will be the template for drilling the tap holes in the mounting base, so you should build the base and fit the casting to it before you drive it onto the column.

Clamp the casting to the assembled base, drill one 1/4" hole in the base through one of the holes in the casting, separate the parts and tap the base 5/16"-18. Enlarge the hole in the casting to 5/16", and bolt it to the base to drill the remaining tap holes in the base assembly. Then tap the remaining holes, enlarge holes in the casting, and make a trial assembly with all four bolts.

When the casting is fit to the base you can drive it onto either end of the column. Rest the column on a block of hardwood, and use another block to protect the casting as you drive it on with a one or one and a half pound hammer.

Finally, bolt the column base casting to the mounting base with four 5/16"-18 X 1" cap screws with lock washers.

THE MOUNTING BASE

My original plan was to present the design with a cast aluminum base. After casting it and doing preliminary work on it, I realized that it was the machining of the casting that would be a serious problem in the developing shop. The fabricated mounting base is the more practical idea at this point.

The entire assembly is shown as bolted together with 1/4"-20 X 1/2" cap screws with lock washers, but it can be welded if you have the equipment.

I used 1/4" X 3" cold rolled steel for the top, but hot rolled steel would work almost as well.

Dimensions are not critical, but of course the slot should be carefully spaced, the assembly should be well squared, and it must rest firmly on a flat surface so it won't be twisted when bolted to the bench top.

As in all stacked assemblies using rows of screws to join them, install one bolt completely before drilling the remaining holes.

Only the front member of the frame is notched to admit the head of a carriage bolt for mounting work on the base.

Clamp the top plates to the front and back angles with a length of 3/8" key stock between them to space for the slot. Then drill tap size holes through both members to begin construction. Tap the holes in the top members, and enlarge the holes in the angles to 1/4". Install the cap screws through the bottom.

Square up the assembly as you locate and install the screws through the front and back angles, then instsll the side angles parallel to each other.

Install one of the mounting feet completely, but install the other with just the center screw. Then you can rest it on a flat surface to align it before you install the other two screws.

You can cut the notch in the front angle with a hack saw, then shear away the material between the cuts with a cold chisel while holding the angle in the bench vise.

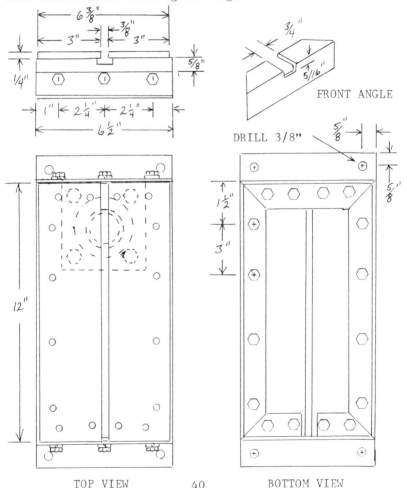

FRONT ANGLE

DRILL 3/8"

TOP VIEW   BOTTOM VIEW

All frame members shown are 1/8" X 1 1/4" angle iron. The locations for the screws are approximate, and all of the screws are 1/4"-20 X 1/2" cap screws with lock washers. Add one or more flat washers if the screws protrude.

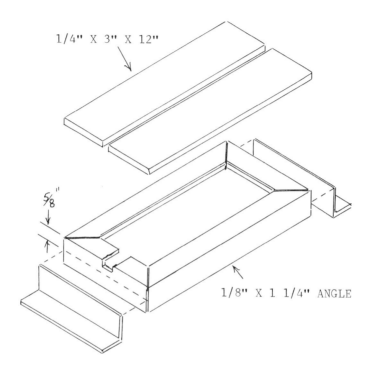

The frame members are cut at 45 degrees, but the fit at the corners is not critical except for appearance.

It would be best to bolt the top plates, but the frame members can as well be welded together.

When the top plates are installed, you can clamp the column base casting in place and drill the 1/4" tap holes through the base casting. Tap the holes in the top plate, enlarge the holes in the casting to 5/16", and drive the casting onto the column before you bolt it in place.

Because the column base casting was bored perpendicular to the bottom surface, the column will be perpendicular to the mounting base top surface. From this point both the column and the base top surface will be reference points for checking the accuracy as components are added.

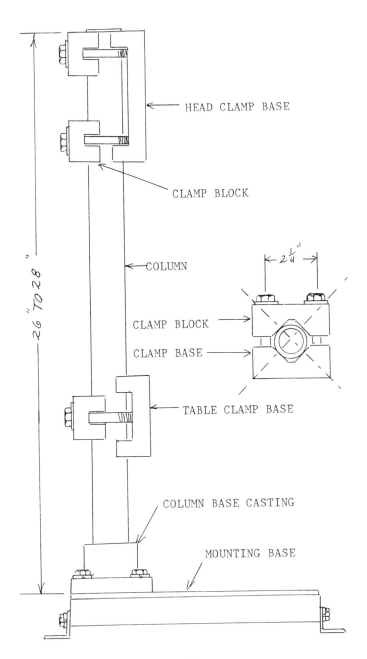
42

THE HEAD CLAMP AND WORK TABLE CLAMP ASSEMBLIES

It would be great if we could cast a one piece head, which is quite a chore in itself. Then all we would need is about $50,000.00 Worth of equipment to machine it and the job would be done. Since $200.00 will buy a pretty good drill press, and most of us don't have $50,000.00 to spend, we'll just forget about that idea and say no more about it. There is a cheaper way.

The clamps can't be made with a circular bore because the column is not of uniform diameter over its full length. The V block clamps make it possible to use a less than perfect column because they make contact on two pairs of opposing points.

The dimensions given are for the 1 5/8" diameter of a 1 1/4" pipe column. If you use a larger column you must widen the base of the V, so clamping pressure will be on opposite sides to avoid distorting the column.

The head clamp and the table clamp differ only in the length.   All three clamp blocks are identical, so only three simple patterns are required.

Only a minimum of draft is required on all three patterns. Wipe a very small fillet on all inside corners.

Use 3/4" thick pine stock, and join the parts with glue and brads.   Form the V's carefully to reduce finishing work on the castings.

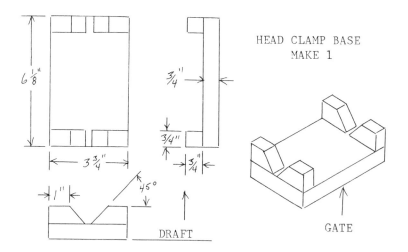

HEAD CLAMP BASE
MAKE 1

43

The angular blocks are all identical, and it requires twelve of them to make the three patterns.

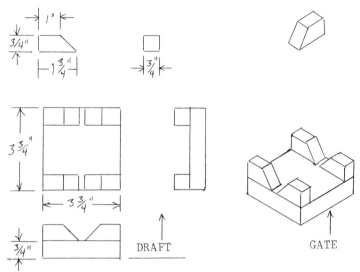

TABLE CLAMP BASE, MAKE 1

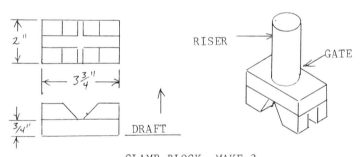

CLAMP BLOCK, MAKE 3

Make just one of each pattern, but three castings are made with the clamp block pattern.

All require a 1 1/4" sprue on either side as indicated, and a 1 1/4" riser connected to the opposite side will be needed to avoid a shrink cavity. You could also set the riser in the center of the pattern, as indicated on the clamp block pattern drawing.

As always, clean up the cavities in the mold carefully before closing up to pour.

MACHINING THE CLAMP BASES

The drawing on page 43 shows a top view of the assembled clamps. Two 3/8"-16 X 3 1/2" cap screws are used with each clamp block.
Note the positioning of the blocks on the side view. Drill 5/16" holes through the blocks on 2 1/4" centers, and clamp the block and base together with a length of pipe between them to guide the drill as you drill the tap holes in the base. Enlarge the holes in the blocks to 3/8", and tap the holes in the bases 3/8"-16. Make sure the V blocks contact the column evenly, and correct to a good fit by filing if neccessary.
By bolting the clamp bases to the face plate with a short length of pipe the same diameter as the column for a spacer and facing it off, the machined surface will be parallel to the column. See the photos of the table clamp base on pages 23 and 24.
The head clamp base requires only facing off, while the table clamp base needs to be drilled and tapped. You can use diagonally opposite holes in the head clamp base for mounting, but it will probably be best to drill and tap two holes near the center just for mounting on the face plate. This casting is just barely wlthin the capacity of the home made lathe, so it must be carefully centered on the face plate. Use cap screws and flat washers through the slotted holes in the face plate for mounting, and adjust so that the flat surface is parallel to the face plate. Tighten the bolts just enough for a firm hold, but don't over-tighten and distort the face plate or the pipe. Rotate the work by hand, to be sure of clearance, and file the corners of the casting if it won't clear when centered.
Mark the center of the table clamp base and punch it. Mount it on the face plate with the length of pipe, and bring up the tail center to enter the punch mark as you center it on the face plate. Center it carefully and adjust it parallel to the face plate before facing it off.
Face it off clean, and cut a small dimple in the center for the drill to start. Use the tail stock chuck to drill a 5/16" hole through the center, and guide a 3/8"-16 tap with the tail center to tap the hole. Turn the face plate by hand as you tap the hole. Face off a depression about .020" deep to a diameter of about 2 1/2" so that only the outer diameter of the table arm will contact it. The remainder of the work table will come a bit later.

THE HEAD CHANNEL CASTINGS

This is a simple pattern with even 1/2" section thickness throughout. There are two identical channels, so only one pattern is required. Only a minimum of draft is needed, and a very small fillet on the inside corners. A routine mold, fed with a 1 1/4" sprue at either end.

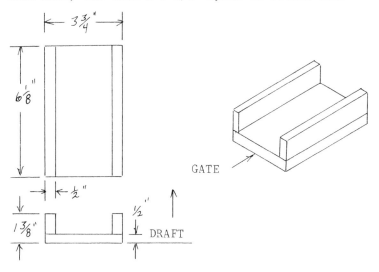

MACHINING THE HEAD CHANNEL CASTINGS

The head clamp base is installed on the column, and its surface has been machined parallel to the column. The head channels will be machined with their front and back surfaces parallel to provide a true vertical surface for the quill guide when they are bolted to the head clamp.

The beginning procedure is the same as for machining the accessory angle plates. A pair of 5/16"-18 holes are tapped in one leg of each channel, the corners are filed so it will rest on the face plate without rocking, and one channel is bolted to the face plate with shims to bring it to right angles to the face plate. See the photo on page 15. Then, using an accurate 3/4" wide guage, such as the blade of a small square or a rule, the second channel is bolted to the face plate and shimmed parallel to the first. The exposed legs are faced off smooth, then the channels are inverted and bolted to the face plate without shims to face off the second side. The result is a pair of channels

46

that are exactly parallel. This is another job that is just barely within the capacity of the lathe, so you may have to file the corners a bit if they don't clear the bed when centered on the face plate.

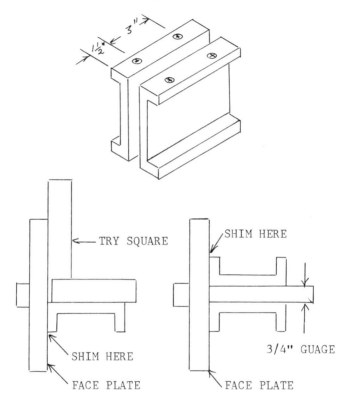

Take plenty of time to set up accurately, and make a series of passes to clean up the surface. Only a light cut can be made because the work is so far from the support of the spindle bearings.

The photo on page 48 shows the first side being faced off. Note the shims that were cut from an aluminum beverage can.

It requires two bolts in each leg for the first set up because the mounting surface is rough. Only one hole is tapped in the center of each of the faced off legs for the second set up, because the machined surface will rest and grip well on the face plate without the shims.

INSTALLING THE HEAD CHANNELS

The three holes that are tapped in each channel were for mounting on the face plate only. They will have no further use in the construction. The leg that has just one hole will be bolted to the head clamp base with two new holes, to be drilled through the clamp base into the channel.

Install the head clamp at the lower end of the column and use an accurate square to scribe a vertical line that is exactly perpendicular to the mounting base top surface. Then remove the base from the column and scribe a line on each side of the center, 3/8" away and parallel. This is the guide line for locating the channels. Align the channels carefully on the scribed lines, clamp them securely, and drill 1/4" holes through both the base and the channels from the rear side of the base.

The holes in the channels will be enlarged to 5/16", and the holes in the base are tapped 5/16"-18.

Bolt the channels to the base with 5/16"-18 X 1" cap screws with lock washers, and install the clamp at the top of the column. You now have a true vertical surface on which to mount the quill guide.

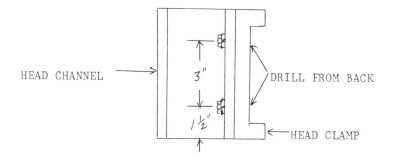

HEAD CHANNEL → 3" ← DRILL FROM BACK
1½"
← HEAD CLAMP

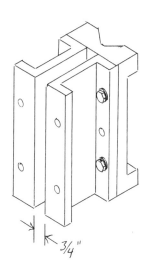

¾"

 These methods and standards produce something less than a perfect assembly, but it will be so close as to be difficult to detect the error if you have worked carefully. Because the column is not fully machined, it would not do any good to check alignment with a dial indicator because no reliable reading could be had. Either the head or work table could appear perfectly aligned on one portion of the column, and possibly be out of alignment by a half degree on another position. More practical tests can be made at a later time, and there are simple ways to correct errors.

CHAPTER IV

THE QUILL FEED MECHANISM

 The spindle is mounted in the quill, which slides in its close fitting guide to feed the drill accurately and with firm support.
 Since milling a rack and cross boring for a pinion is out of the question for most of us, we need a device for advancing the quill that is within our ability.
 Many have solved the problem with a roller chain and a small sprocket, and the idea works very well. Its cost is moderate, and it works nearly as well as a rack and pinion. A bicycle chain will work, but a standard # 35 roller chain is better.
 I'm still hung up on the idea of doing a ten dollar job for fifty cents, so my solution is a cable winch. I used a 1/16" cable that was intended to operate the brakes on a bicycle. It cost $1.00 and I used just half of it, so I was able to stay within the proposed budget.
 There is no doubt that a chain drive is much stronger, and that should be your choice if you plan on a large amount of heavy duty work with your drill press, but the cable winch will serve for all ordinary purposes.
 A small sprocket would be installed on the cross shaft in place of the winch drum, and the stud requires a bracket to attach it to the chain. All other details of construction remain the same. The smooth top idler will work as well for the chain as it does for the cable.
 The secret of success with the cable feed is to position the drum so that the cable enters and leaves the hole so that it does not flex at the sharp bend. It will give long service if properly installed, but will soon break if allowed to flex each time the quill is fed.

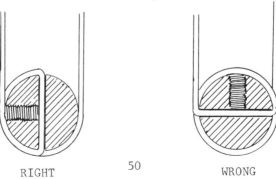

RIGHT   WRONG

THE CHAIN DRIVE QUILL FEED

The cross shaft is fitted with a 10 tooth #35 sprocket, and the idler is positioned so that the front loop of the chain travels parallel to the vertical surface of the head channels. The 3/8" stud is threaded into a tapped hole in a 2" length of 1/4" X 3/4" steel and locked with a jamb nut. The stud bracket is fastened to the chain with two 3/16" bolts through the chain and into a tapped 1/8" X 1/2" X 1" steel nut plate. A short length of cable is fastened to the stud bracket, runs over the idler, and connects to the return spring.

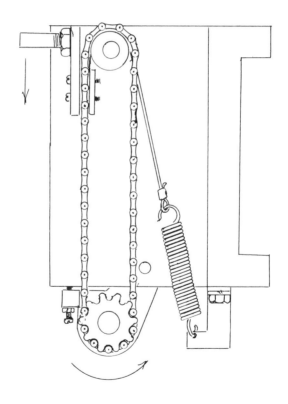

CUTAWAY VIEW WITH RIGHT CHANNEL CLEARED AWAY

THE CABLE WINCH FEED

    A special cross shaft is prepared, and the stud is a different arrangement, but the cable drive is the same in all other details.

    I've used the cable method some years ago on a large floor model drill press. At the time I used 1/8" cable, and it was quite a chore to thread it. It would not be practical to use a 1/8" cable in a small machine, but it will greatly strengthen the feed if you use a double strand of 1/16" cable. That would require larger holes in the drum and stud, but all other details would be the same as for the single strand. I've drilled numerous holes with my single strand feed, including 5/8" holes in steel, and it appears amply strong for the work.

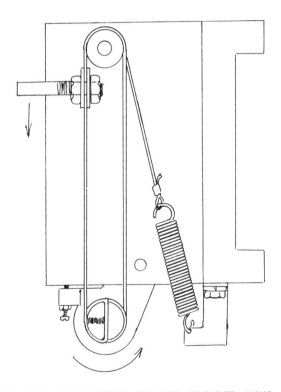

CUTAWAY VIEW WITH RIGHT CHANNEL CLEARED AWAY

The stud is a 3/8" cap screw with the head cut away, and a 1/8" hole is drilled through 3/8" from the threaded end. A pair of nuts drive two flat washers together to clamp the cable.

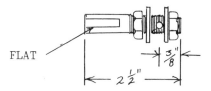

FLAT

The 2 1/2" length is greater than needed. The actual length will be determined when the quill is installed. A small flat is ground or filed on the top surface for a set screw to seat against.

Threading the cable is easy if done in sequence, but it requires a bit of study. A single strand is illustrated, but the procedure is the same with a double strand.

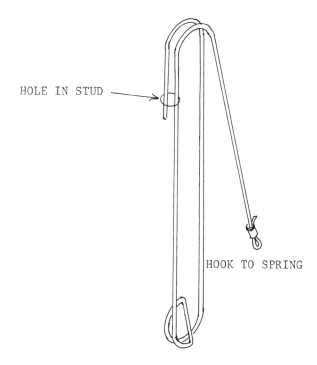

HOLE IN STUD

HOOK TO SPRING

CABLE THREADING SEQUENCE

If you can't find a small sleeve to make the spring loop, you can drill a 1/8" hole in a short piece of steel rod. Make the loop as small as possible, and mash the sleeve flat with a hammer.

Hook the loop to the return spring and pass the cable over the idler.

Slip the free end through the hole in the stud and then through the hole in the drum.

Rotate the drum so that the cable enters at the bottom and leaves at the top, and clamp the cross shaft so it can't change position.

Draw the cable snug so that the return spring stretches no more than 1/4", and lock the set screw in the drum.

Pass the free end of the cable over the front of the drum, under it and up the rear side, then over the top of the idler and through the hole in the stud.

Draw the cable snug to eliminate all slack, then raise the stud assembly to the top and tighten it up to lock the cable. The stud should be locked as high as possible.

The adjustment screws on the cross shaft bearings will remove any slight amount of slack to eliminate back lash.

You'll notice in the photo that I failed to rotate the drum so that the cable entered at the bottom. This caused the cable to flex at the sharp bend each time the quill was advanced, and it wasn't long before it began to fray. I did the job right the second time, and it has endured hundreds of quill advances with no sign of failure.

With these principles understood you can choose either chain or cable feed, and proceed to make the parts.

THE CROSS SHAFT BEARING SUPPORT CASTINGS

One of the most difficult jobs in machining a one piece head would be boring for the cross shaft bearings. By casting them separately and machining them on the face plate, the job becomes simple.

These castings will be the same whether you use chain or cable feed.

An opposing pair is needed, but both castings are made from one three piece split pattern. The main body can be a bit tricky if you are not experienced in layout, so follow the simple steps and it will be a breeze.

Prepare a block of pine 3/8" X 2 1/4" X 3 1/4", and draw a line on the center. Then draw two crossed lines as shown on the sketch. The crossed lines are the centers for the circular portions.

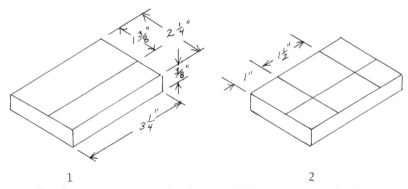

1   2

Use a compass or a circle template to draw circles on the centers made by the crossed lines.

Only one of these is required, and it will be formed without any draft because it is reversible.

This is a good exercise in pattern making, molding and casting, and it's easy and fun to machine. It will be a real " show off " part of your machine.

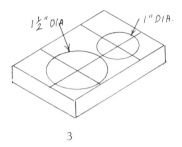
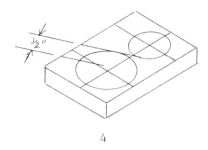

3   4

When the circles are drawn, draw a diagonal line tangent to them, and another line parallel to the tangent and 1/2" into the large circle.

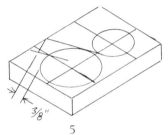
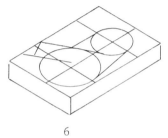

5   6

Connect the parallel lines at right angles 3/8" from the large circle. Then draw a line tangent to the small circle and parallel to the center line. Finally, draw a line tangent to both circles to complete the outline.

Cut it out to the line with a coping saw or jig saw, and sand the edges smooth. There is no draft, so the edges must be exactly perpendicular and very smooth.

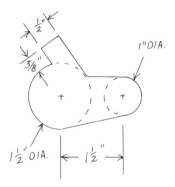

The hub is a simple disc of pine with minimum draft. It is aligned with the main body by a small pin in the exact center. A small hole is drilled in the top to accept a guide pin in the end of the sprue pin.

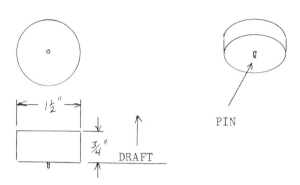

PIN

The lug for the adjustment screw is a simple block with minimum draft, and it is aligned with the main body by two small pins.

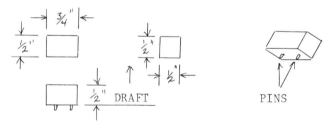

PINS

The pin holes in the main body are drilled through at exact tight angles to the surface so that the loose parts can be set on either side to make an opposing pair. The pins can be small nails, and you can use a nail of the same size to drill the holes in the main body. Make the pins very short, cement them firmly in the loose pieces, and make them fit the holes in the main body freely so they will separate easily when the mold is opened.

Prepare a 1 1/4" sprue pin with a small guide pin in the end to fit the hole in the top of the hub piece.

The castings will be bored and fit with bushings, and they will be bolted to the lower end of the head channels.

The hub will face to the outside on either side, and the adjustment screw lug will face to the inside.

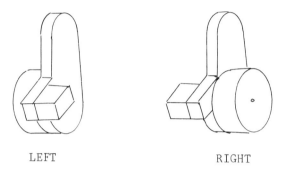

LEFT   RIGHT

MOLDING THE CROSS SHAFT BEARIGN SUPPORTS

Lay the main body pattern on the molding board and set the adjusting screw lug pattern in place.

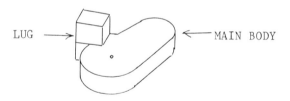

Ram up and vent the drag, rub in a bottom board, and roll over. Set the hub in place and set the sprue pin in place on the hub.

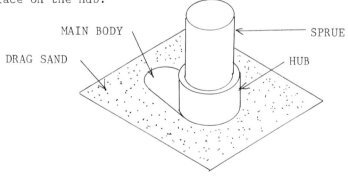

Ram up the cope carefully to avoid striking the sprue pin. Finish the sprue opening to the funnel shape and pull out the sprue pin.   Rub in a bottom board and open the mold, laying the cope down on the bottom board. Swab, rap and draw the hub pattern from the cope, and likewise the main body pattern from the drag. You should be able to grasp one of the pins on the lug pattern to draw it from the drag. Clean up carefully, especially the sprue, and swab any areas that threaten trouble. Close up and pour rapidly. The sprue serves as a riser, or it might be more properly calles a " sinking head ". It supplies molten metal to the heavy hub section as it shrinks on solidification.

Repeat the mold beginning with the lug on the opposite side of the main body for the other bearing support.

MACHINING THE CROSS SHAFT BEARING SUPPORTS

Saw off the sprue at the shoulder and locate the center of both circular portions. Drill a 1/4" hole in the small end, which is the mounting ear, and tap it 5/16"-18. Make a punch mark in the center of the hub.

Mount the casting to the face plate with a set screw collar and a couple of washers as spacers so the lug will clear the surface of the face plate.

Bring up the tail center to enter the punch mark in the hub as you adjust it to center.

Install a couple of set screw collars as a stop lug on the face plate to prevent the casting from slipping as you drill and bore the hub.

Face off the end of the hub and cut a small cone center to start the drill, and use the tail stock drill to drill a 3/8" or 1/2" starting hole.

Carefully bore to .625" for a snug push fit on a 5/8" X 1/2" bronze bushing.

Push the bushing in and repeat the procedure for the opposite support.

There is no real need to machine the outside of the hub except to improve its appearance if you wish.

Drill and tap a #10-24 hole in the center of the adjusting screw lug so that the screw will engage the bottom of the head channel when the lug is installed.

A #10-24 X 1" machine screw with a jamb nut serves as the adjustment screw.

Note the set screw collar stop lug bolted to the face plate to prevent the casting from slipping.

BORE TO .625" ALL THE WAY THROUGH THE HUB

INSTALLING THE CROSS SHAFT BEARINGS

As no steps were taken to make the bearing supports identical, one side is installed and used as a guide for positioning the opposite side.

Locate, punch and drill a 1/4" hole 3/8" from the bottom of the head channel and 1" from the rear leg, on one side only.

Enlarge the tapped hole in the ear to 5/16" by drilling out the threads. Don't enlarge the opposite side at this time.

Tap the hole in the head channel 5/16"-18, and bolt the bearing support to the channel with a 5/16"-18 X 1 1/4" cap screw. Install a jamb nut on the inside.

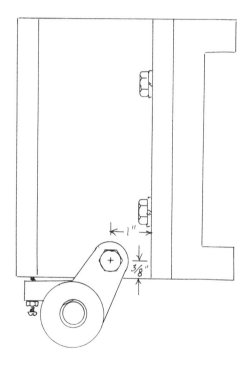

Slip a short length of 1/2" shaft through the bushing and slide the opposite bearing support onto the shaft and against the channel. Clamp it to the channel as you adjust its position, and drill a 1/4" hole in the channel through the tapped hole in the ear.

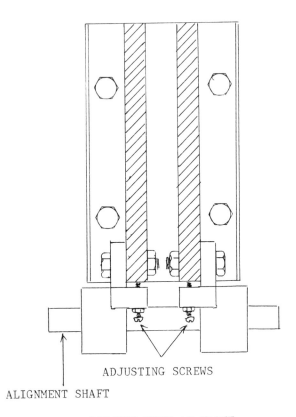

ALIGNMENT SHAFT   ADJUSTING SCREWS

CUTAWAY VIEW AT FRONT

Tap the second hole in the channel and enlarge the hole in the ear to 5/16".

The 1 1/4" bolts are used because they have a shoulder that is not threaded. A standard 1" bolt will have threads all the way to the head, and would not make a good pivot for the bearing supports. Cut a portion of the end of the bolts away so there will be clearance for the cable and return spring.

THE CROSS SHAFT

If you elect the chain drive feed, the cross shaft will simply be a 6 1/2" length of 1/2" cold rolled steel round. The sprocket will be locked to the shaft with its set screw to replace the winch drum. A flat should be ground or filed on the shaft for the set screw to seat against.

A special shaft is machined on the lathe if you choose the cable winch feed. It's a simple between centers job, and the rough stock is a 6 1/2" length of 1" or 1 1/8" cold rolled steel round.

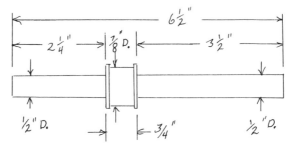

Turn the 1/2" diameter portion to near size and finish by filing and polishing with emery. The shoulder of the drum is an aid in installing the cable, but it has no function in the operation of the feed. It's size can be any diameter that is left after cleaning up the O. D., and about 1/16" wide.

When the shaft is machined to size and shape, drill a 1/16" hole through the center of the drum, and countersink it slightly on each end to break the sharp edge.

CUTAWAY VIEW OF DRUM

Drill a 7/32" hole at right angles to meet the 1/16" hole, but not beyond the center. Tap the hole for a 1/4"-28 set screw.

Because the set screw would cut the cable, a 3/16" ball bearing is dropped in the hole and a socket head set screw drives the ball against the cable to lock it.

THE IDLER PULLEY

This can be the same size and shape whether you use a chain or cable feed.

I mounted a short length of 1" cold rolled round in the V block angle plate, and bored a 1/2" hole about 1" deep. Then I hacksawed the bored blank from the stock, installed a 1/4" set screw, and mounted it on an arbor between centers to machine it to the same size and shape as the drum. The set screw is used only for machining, and a 1/2" X 3/8" bronze bushing is installed in the bore to complete the idler.

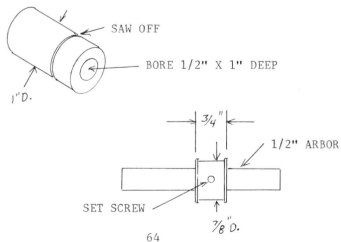

INSTALLING THE CROSS SHAFT AND IDLER

One cross shaft bearing support must be removed to install the cross shaft. The longest end extends to the right. Install a set screw collar on the left end to position the drum midway between the channels. A flat should be filed or ground on the shaft for the set screw to seat against. Lightly lock a set screw collar on the right end to hold the drum in center position.

A 3/8"-16 X 2 1/2" cap screw forms the axle for the idler pulley, and it's locked with a jamb nut. The right hand channel will be drilled 3/8", and the left hand will be drilled 5/16" and tapped 3/8"-16. The best way is to step drill both holes to 5/16", enlarge the hole in the right channel to 3/8", and tap the hole in the left channel using the right hand hole to guide the tap. These hole centers must be carefully located so the cable or the chain will run parallel to the front of the channel. We can view the head as though the left hand channel were removed while we discuss locating the hole centers.

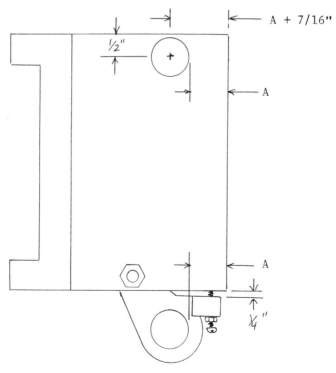

As the center of the bearing support moves down and to the rear when the adjusting screw is turned in, it should be adjusted to its estimated final position now. A gap of about 1/4" will be close enough.

With the drum in position and the adjusting screws set, measure the distance from the front surface of the channel to the front surface of the drum. Add one half the diameter of the idler to the " A " dimension and locate the idler center 1/2" below the top edge of the channel. This will bring the front surface of the idler to the same distance from the front as the drum. In the case of the cable feed it will be " A " plus 7/16", but you will have to calculate the root diameter of the sprocket to find the " A " dimension if you use the chain feed. A slight error will not be serious, but get it as close as possible.

A small adjustable square is a handy tool for such a job because it works nicely as a depth guage. With the vertical center line scribed on the top of the channel, it is easy to use the square to drop the line down the outside of the channel and then scribe a cross line at a depth of 1/2".

Drill a 1/8" hole in each channel. Then drill a 3/16" hole through both channels from one side. Follow with a 1/4" drill, then a 5/16" drill. Enlarge the right hand hole to 3/8", and tap the left hand hole 3/8"-16.

Install the idler with its axle and jamb nut, and turn the adjusting screws out so that the lugs rest on the bottom of the channel.

THE RETURN SPRING AND BRACKET

The spring bracket is made of a 1" length of 1/8" X 1 1/2" angle iron, and it's bolted to the bottom of the head clamp base with a 1/4"-20 cap screw tapped into the base. See the drawing on page 52.

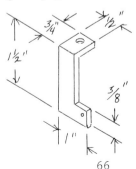

I'd like to give an exact specification for the spring, but this is one of the hazards of working out of the junk box. The one I used is wound of 1/16" wire, and it measures 7/16" in diameter and 3" long, including the loops. There is a difference in tension in springs of the same apparent size, so you may have to do some experimenting to find the right tension. Stiff enough to return the quill, but not so stiff as to make the feed difficult.

INSTALL THE CABLE OR THE CHAIN

Install either the cable or chain as discussed earlier, and adjust the cross shaft bearing supports to take up any slack. Either the cable or chain is installed with the adjusting screws withdrawn so that a minimum of travel is used to take up slack. Adjust for a smooth motion without tension between the drum and idler.

THE HAND WHEEL SPOKES

Wooden macrame beads are available in a 1" diameter with a 3/8" hole through the center, and this makes a perfect pattern for the ball end of the spokes. Get three of them so you cam mold all three spokes in a single flask.

Glue a 3/8" length of 3/8" dowel in one end of the hole to plug it, and smooth up the plugged end with body putty.

A 6" length of 3/8" steel rod is slipped into the hole to complete the pattern, and the same rod is used as a core when the mold is poured.

A sand match or a plaster match would be used if this job was to be done repeatedly, but such labor is not justified for a one time job. Simple bedding is the right way.

Just ram up a blank drag over a smooth molding board and roll it over.

Press all three patterns into the parting face of the drag to the half way point.

Set a 1 1/4" sprue pin about 1" away from the center ball and ram up the cope.

Remove the sprue pin, finish the sprue opening to a funnel shape and open the mold.

Lift out the patterns, and run the vent wire through the mold from the inside of each cavity in both the cope and the drag.

Cut a gate from each ball cavity in the drag to the print left by the sprue pin, and clean up the mold.

Slip the beads off the rods, and set the rods in the original print that they left in the mold.
Close up the mold and pour.

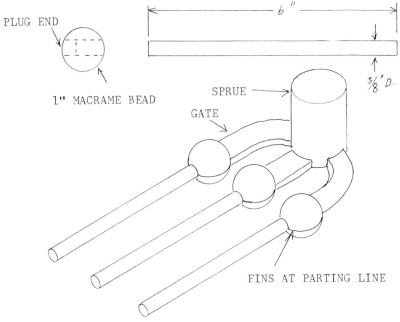

Cut off the gates, file up the rough areas at the fins and the gate stub, and chuck the rod in your electric drill to polish up the ball with emery cloth.
Cut 3/8"-16 threads for about 1/2" on each spoke.

THE HAND WHEEL HUB

A simple pattern made of two 2 1/8" discs of pine that are glued together, and a 1 1/4" X 3 1/2" sprue pin with a small alignment pin in the end to enter a matching hole in the hub, and the sprue forms a shank so it can be mounted in the V block angle plate for the boring operation.

The same pattern is used to mold the lower bearing support for the quill, so you may as well cast two of them.

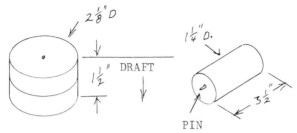
PIN

Only minimum draft is required on the disc portion, and you can build a mound of sand on ahe top of the cope if your flask is not deep enough to produce the full length shank that is needed for mounting on the lathe.

The photo above shows the quill casting being bored to a 1" diameter for a depth of about 1 3/4". Notice the pulley and collar that are used for counter weight so a higher speed can be used.

The hand wheel hub is bored in the same way, but its bore is just 1/2" to fit the cross shaft.

Bore it carefully for a good fit on the sross shaft, and hack saw it from the shank. Install two 1/4"-28 set screws at 90 degree intervals 3/8" from either end of the casting, and mount the bored casting on a 1/2" arbor between centers for outside machining.

Reduce the outside diameter to 2" over the entire hub, then reduce the set screw end to 1 1/4" for a shoulder of 5/8". Machine a 25 to 30 degree bevel beginning 3/8" from the opposite end. Face off both ends to clean them up.

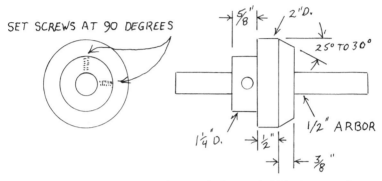

It would be nearly impossible to drill the angular tap holes for the spokes free hand, so a simple jig is used to mount the hub on the face plate at the proper angle.

A power lawn mower axle is an old friend by now if you have built the metal lathe. It provides the arbor for the jig, and its hex head is a convenient index to space the tap holes at 120 degree intervals.

The mounting block is a simple aluminum casting with a 3/8"-16 hole tapped in the base for mounting, and another of the same size tapped in the angular surface for the arbor. I machined mine from a scrap by mounting it in the face plate angle clamps that are described in books 3 and 4, but I think it easier to make a pattern of the correct size and shape and finish with a file. If you have the face plate angle clamps, use them to hold the base as you drill the tap holes with the tail stock drill. They should be perpendicular to the surface for an accurate set up.

The angular face is at the same bevel as the hub so it will be supported at right angles to the turning axis of the lathe.

The lawn mower axle has a 1/2" shank about 1 1/2" long, and its threaded portion is 3/8"-16 about 1/2" long.

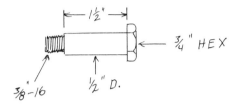

With a line scribed on the fixture that is parallel to the axis of the arbor, it is easy to bolt the fixture to the face plate and align it horizontally with a tool in the tool post as you center the beveled area of the hub with the tail stock center.

Lock the hub on the arbor with either of the set screws in the hub, and scribe a mark to align with any point on the hex head. Cut a small dimple to start the drill, and drill a 5/16" hole with the tail stock chuck. Use the tail stock chuck to hold the tap as you start the threads so they will be true with the axis. Rotate the face plate by hand while tapping.

Loosen the set screw and advance the mark two points on the hex head of the arbor, and lock it for the second hole. Drill and tap it and repeat for the third hole.

One or the other of the set screws will be inaccessible for at least one of the set ups, but either one is sufficient.

Screw the spokes into the hub, file or grind flats on the cross shaft for the set screws to seat against, and install the hand wheel.

Prentice 12 Inch Swing Upright Drill.

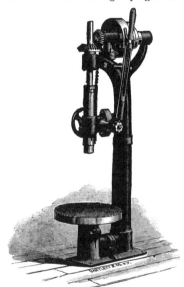

The engraving is from HILL & Clarke's 1885 Catalog. This was a bench model, but not a toy. It weighed 225 pounds, and its mounting base was 12" X 24". A #2 morse taper socket in the spindle, and a speed range low enough for large hole drilling made it a heavy duty machine. The price, including an overhead countershaft, was $85.00.

Note how the components were cast and machined separately, and bolted together with flanged joints. They faced the same problem we do with limited facilities.

The chain at the top of the quill runs over pulleys, and supports a weight to counterbalance the quill.

CHAPTER V

THE QUILL

    I used 5"-4"-3"-2" die cast step cone pulleys with a 5/8" bore, and was able to find them only in Sears Tool catalog. You'll need two of them, and if you use the reduction for large hole drilling you'll need a third pulley which can be a 4"-3"-2" bored to fit your motor shaft.
    There are two 5/8" ball bearings in the quill, and a pair of the same size are used in the reduction. The driven pulley uses two 1" ball bearings. I purchased all of my bearings, bushings and set screw collars from the W. W. Grainger Company, who have branches all over the country. This is a wholesale company, and you must buy through a dealer or your employer if you are not a re-seller or manufacturer. Grainger's number for the 5/8" ball bearing is R-10-ZZ-1L006, and it interchanges with a Fafnir 57KDD. It is an electric motor grade bearing with 5/8" bore and 1.375" O.D., and its thickness is 5/16". The 1" bearing

is a Grainger R-16-ZZ-1L009, and it interchanges with a Fafnir S10KDD. The bore is 1", O.D. is 2", and it's 1/2" thick. Grainger's number 2X529 pillow block has a 5/8" X 7/8" flanged bronze bushing that serves well for the thrust bearing, and they stock finished bore bronze bushings for the cross shaft bearings and the idler pulley.

Lacking thread cutting gear on the lathe, it is near impossible to cut threads on the spindle that would be accurate enough for mounting the chuck. The easiest mount would be a chuck with its back bored 5/8" and fit with a set screw. You can get one from Gilliom Mfg. Inc., St. Charles, Mo. 63301. You will see Gilliom adds in magazines offering metal parts kits for wood working machinery. I recommend their kits highly, for I have built some of them to my complete satisfaction.

I purchased the chuck for my drill press from Wholesale Tool Company. It's a 1/2" key tighten chuck with its back bored for a #33 Jacobs taper. It comes with a #1 Morse taper adapter shank, so it can do double duty as a tail stock chuck for the lathe.

If you are going to be doing metal work you really need to know about the Wholesale Tool Company. Send them $1.00 for a catalog and expect to be pleased. They have four large outlets, so order from the nearest one.

Wholesale Tool Co. Inc., 12155 Stephens Dr., Box 68, Warren Mich. 48090.

W T Tool/North, Inc., 1234 Washington St., Box 481 Stoughton, Mass., 02072.

W T Tool/South, Inc., 4200 Barringer Dr. Box 240965, Charlotte, N. C. 28210.

Wholesale Tool Co. of Oklahoma, Inc., 7240 E. 46th St. Box 45952, Tulsa Oklahoma, 74145.

All branches have a toll free ordering number, and they accept bank credit cards. Their minimum order amount is $15.00, and they sell to individuals like you and I. A really great stock at good prices, and service is fast.

These are not advertisements from any of the companies mentioned.

All other needed items should be available in most moderately well stocked hardware stores and farm supply stores.

Check your junk box and any broken down appliances you find laying about. You might find some of what you need at no cost. A burned out 1/2" electric hand drill may have a chuck with a tapered back bore if it was a reversible model. See if you can't find the bearings, bushings etc..

THE QUILL SUPPORT CASTINGS

The lower bearing support is identical to the hand wheel hub, and you should have it cast already.

The upper support is longer, and it has a lug cast on that will be drilled to receive the feed stud.

Three 2" diameter discs of 3/4" pine stock make up the main body, and the tapered lug is fastened to the side with glue and brads. The mold is fed with the same sprue that was used for the hand wheel hub and the upper support. It will be best to make the circular part a bit oversize and mount it between centers to true it up so it will be true round and concentric with the guide pin hole for the sprue. Add the tapered lug to the finished body and wipe a smooth fillet at the junction.

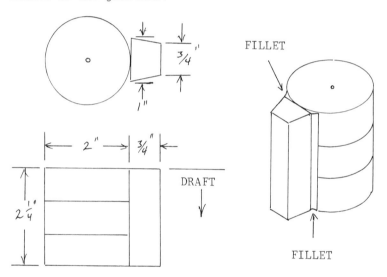

Only minimum draft is required, and note that it is parted at the same plane that is used for the sprue guide pin. Otherwise you could not draw it from the mold.

The mold is fed with the 1 1/4" sprue which provides the shank for mounting it in the V block angle plate.

MACHINING THE BEARING SUPPORTS

Both castings are bored to 1" for mounting on an arbor, and they are counterbored to 1.375" to recieve the ball bearings. The bores must be concentric, so they are done in sequence without changing the original set up.

Mark the center of the casting and bring up the tail center to position it on the turning axis of the lathe as you set up for boring. Drill a 3/8" or 1/2" starting hole with the tail stock drill to a depth that reaches about 1/2" into the shank. This will be about 2" on the lower support, and about 2 1/2" on the upper support.

1 1/4" of the lug is cut away with a hack saw so the diameter can be reduced to enter the quill sleeve. This is a much easier job than to make a more comples pattern and mold with a loose piece.

Bore all the way to the bottom to 1" diameter, and use a short length of 1" shaft as a guage to be sure it will fit on the 1" arbor. Then counterbore to 1.375" to a depth of 5/8".

These operations take considerable time, and you must be very patient and measure with great care. If you are new to this work, it may take three evenings or more to do the job, but remember that you are gaining skill that will repay you for your patience later. It is no great loss if you spoil a casting and have to start over, and you will do better on the second try. Don't give up.

The bearings themselves cas serve as the gauge for the 1.375" counterbore, but be careful not to get them stuck in the bore before the job is finished. The casting is not designed for a force fit bearing, because there is no way provided to drive it out of the bore for replacement. It should slide into the bore with finger pressure, and slide out with no more than very light taps on the bearing. It would be better to have the counterbore .001" oversize than .0005" undersize.

```
        5/8"                              5/8"
                                                    
  1.375"    3/16"              1.375"    3/16"    1"
                                                    
                 SET SCREWS
      1"
              LUG CUT AWAY

       UPPER                      LOWER
```

A 1/4"-28 set screw is installed in each casting at a point about 3/16" from the shoulder of the counterbore. The opposite end of each casting will be reduced to fit the inside of the quill, and the set screw must not enter at the area that is to be machined.

After the boring is completed, and you have tested to be sure that it will fit the arbor and that the bearings will fit properly, cut the shank off with a hack saw and mount the castings on a 1" X 8" arbor between centers for the outside machining operations. The arbor should have a flat filed or ground for the set screws to seat against.

The quill tube is a 5" length of 1 1/4" pipe, and its inside diameter is anything but smooth and true. It will be machined on a pair of collars that will swage the inside diameter to a more definite dimension, which was found by experimentation based on an estimate of the average of measurements. The bearing supports will be shouldered so that they can be driven into the quill tube. 1.438" will be the diameter of the shoulder, and that will be a tight fit when it is driven into the tube. Machine both shoulders, but don't drive them into the tube until it has been machined.

The photo above shows the upper support with its shoulder machined to size, and the end of the lug has been machined lightly to clean it up. The lower support is being reduced to 1.875" after machining the shoulder.

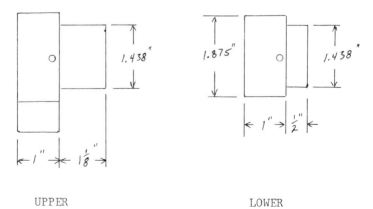

UPPER   LOWER

The upper support is bolted to the angle plate and a 3/8" hole is drilled through the lug to receive the feed stud. Align the casting carefully with the turning axis of the lathe so that the hole is well centered. A 1/4"-28 set screw is installed in the top of the lug to lock the feed stud.

THE QUILL TUBE

A 5 1/4" length of 1 1/4" black pipe furnishes rough stock for the tube. Its nominal outside diameter is not easy to measure because it is so rough, but it is 1.660" if it were perfectly round. There is ample stock to finish smoothly at 1.625" if it is carefully centered on the lathe. Cut the pipe with a hack saw rather than pipe cutters so that the ends will not be reduced.

The inside diameter is extremely rough, but the special arbor will clean it up and swage it to a more definite size.

Standard steel set screw collars of 1" bore measure 1.500" on the outside. Lock a pair of them on a 1" arbor and reduce the diameter to 1.438". These will be driven into each end of the tube so it can be mounted on the arbor.

Remove the set screw from one of the collars and drive

it into either end of the tube. Be careful not to damage the tube or the collar as you drive it in about 3/4".

Drill a 5/16" hole 3/4" from the opposite end of the tube, and drive the second collar in so that its set screw aligns with the drilled hole.

Put the assembly on the arbor and mount it between the lathe centers.

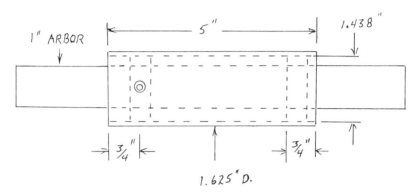

Make light cuts along the entire length of the tube to clean it up, then measure each end to be sure you are not cutting a taper. Adjust the tail stock set over if there is any noticeable difference in diameter at each end of the tube.

Very carefully reduce the diameter of the tube, and face off both ends to reduce the length to 5". The finished size will be 1.625", and it will be best to turn it 1.626" and finish with a file and emery cloth to a smooth surface. Chamfer both ends inside and outside.

When the tube is machined to size, drive out the set screw collars and drive the finished bearing support castings into the ends of the tube. Be very careful not to damage either the castings or the tube as you drive them in. The ends of the tube having been faced off square, and the shoulders of the supports also, the parts will be well aligned when the shoulder meets the end of the tube.

ASSEMBLE THE QUILL AND SPINDLE

A #2X529 pillow block from W. W. Grainger has a 5/8" X 7/8" flanged bronze bushing that will make a good thrust bearing for the lower end of the quill. Their 7/8" set screw collar has a 1.375" outside diameter, which makes it

A perfect adapter to fit the flanged bushing to the lower support counterbore. Just cut the bushing to a shoulder length of 1/2", and push it into the collar.

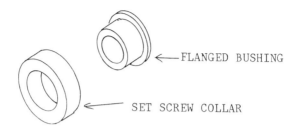

Push the ball bearings into the supports and slip a 16" length of 5/8" cold rolled steel shaft through both bearings. Push the thrust bearing adapter assembly into the lower end against its bearing, and lightly lock a set screw collar at the upper end so that the spindle extends 2" beyond the face of the thrust bearing. Lightly lock the second collar to bear against the thrust bearing.

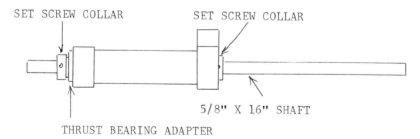

THE QUILL GUIDE CASTINGS

Both castings are made from the same pattern and the ears are cut off one casting to make the cap. This is a simple pattern to make, but take care to align the saddles true square and parallel so the two halves will mate well when joined together. Only minimum draft is needed, but the saddles must be truly perpendicular to the base or it will be difficlt to draw from the mold.

It's a routine mold, fed at either side with a 1 1/4" sprue. Use a 1 1/4" riser at the opposite side to avoid a shrink depression in the base.

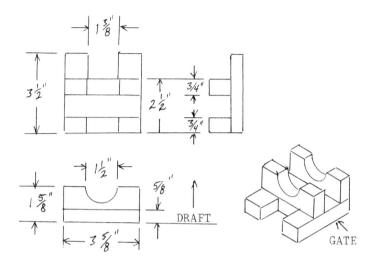

Note that the base is just 5/8" thick, while the saddles are 3/4" thick. This is another casting that is just barely within the capacity of the home made lathe.

MACHINING THE QUILL GUIDE

Saw the ears off one of the castings to form the cap. Drill three 1/4" holes in the base of each casting and tap them 5/16"-18.

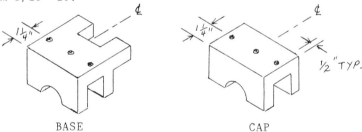

BASE   CAP

The cap will be inverted to mate with the base, and both the broad flat surface and the mating surface must be faced off on both castings. File the mating surface so it will rest on the face plate without rocking, and mount the base casting on the face plate with two 5/16" bolts through the slotted holes in the face plate. Tighten only enough for a good grip, don't distort the plate or casting by over tightening the bolts. Face off the broad flat surface to reduce its thickness to 1/2".

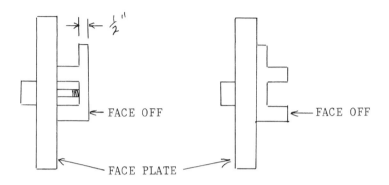

When the flat surface is faced off, invert the casting and mount it with shorter bolts to face off the mating surface. Repeat both operations to finish the cap.

The cap is fastened to the base with four 5/16" bolts, and the holes must be truly perpendicular. This would be very difficult to do free hand, so the job should be done on the face plate with the tail stock drill. The center tapped hole in the cap will serve to mount it on the face plate with the hole centers in line with the turning axis of the lathe. Punch each center midway in the four corners, and bring up the tail center to position them for drilling. Drill a 1/4" hole through each corner.

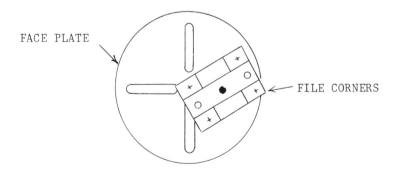

It is likely that you will have to file the corners of the cap so that it will clear the bed. Rotate by hand to check for clearance before you turn on the power to drill the holes. Reposition for each hole carefully, and drill all four 1/4" holes through the casting.

The drilled cap now becomes a jig to guide the hand drill as you drill the tap holes in the base. Assemble the cap and base with one 1/4" bolt through the center of the base and cap, and clamp the assembly in the bench vise to drill the first tap hole. Drill a 1/4" hole through any of the corners, and drill all the way through the base casting. Separate the parts, tap the hole in the base 5/16"-18, and enlarge the hole in the cap to 5/16". Install a 5/16"-18 X 2 1/4" cap screw with flat washer and lock washer to reassemble the parts. Re-clamp it in the vise and drill the second tap hole diagonally opposite. Install the second bolt before you drill the remaining two tap holes.

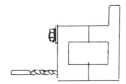

Make up four sets of shims for the mating surfaces so they can be installed before the assembly is bored. Each set should include two .005", two .003", two .002" and three .001". These will provide means to adjust the fit of the quill. Mark one corner of the assembly before you take it apart to install the shims, and put it back together just as it was taken apart.

BORING THE QUILL GUIDE

This is another job that is just barely within the 7" swing capacity of the home made lathe, so set up must be done very carefully.

A single bolt will mount the angle plate near the rim of the face plate, and one bolt through the angle plate and into the center hole in the base casting will nold it as you set up. A second bolt through a scrap of 1/4" X 3/4" steel will clamp the ears of the base to the angle plate when all is well aligned with the turning axis of the lathe.

You can enlarge the hole in the angle plate to 5/16" and use the threads in the base center hole for the first bolt. The 1/4" X 3/4" steel clamp can be drilled 5/16", and the threads in the second hole in the angle plate be used.

The rough bore, as cast, will be a poor reference for the set up, so make a template of stiff paper or sheet metal to tape to the end of the assembly. The joint between the halves should be on the turning axis of the lathe when it is either vertical or horizontal, so the template need only have a single line that coincides with the joint. The sides of the base should be parallel to the turning axis, and you can check it with a try square off the face plate. Give plenty of time and study to this set up, and rotate the work by hand before turning on the power. Add counter weight to the face plate for balance.

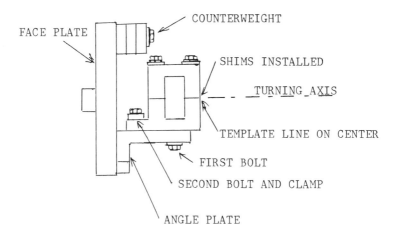

The assembly is faced off on the exposed surface and carefully bored to 1.625" to match the diameter of the quill tube.

Since the faced off base is mounted at right angles to the face plate, it will be bored parallel to the mounting surface of the base. When it is bolted to the head channels it will be parallel to the column.

Be very patient with this job. I spent two evenings setting up and boring because I didn't want to get nearly finished and then have to start over again. Measure twice and cut once is a sound rule to work by.

Notice in the photo below that I have made a heavier 3/8" boring bar, and mounted it on the compound with the clamp dog and a couple of bolts. The same set up is seen on page 17 with a 1/4" boring bar. Either will do the job, but the heavier bar speeds up the work.

INSTALLING THE QUILL GUIDE

Clamp the base of the quill guide in the vise and install the assembled quill in the guide. It is likely that you will have to add or remove shims to bring it to a good fit, so keep notes as you disassemble and reassemble the guide. The tube needs to be well oiled, and you want a very close fit for aligning the guide with the base top.

Mount one of the step cone pulleys on a 5/8" arbor between centers and lightly face off the large end so it can be used to support the spindle exactly perpendicular to the base top surface.

The lower end of the spindle is slipped into the pulley and rested on the base top surface as the head assembly is lowered on the column until its lower edge is 3/4" above the bottom of the quill guide. Lock the pulley set screw lightly, and clamp the pulley to the base top. Do not engage the quill feed stud at this time.

It will be best to remove the hand wheel for clearance during this phase of construction. The remainder of the quill feed mechanism is left in place, but it does not appear in the drawing below for clarity.

When all is well aligned, clamp the ears of the quill guide to the head channels, and drill 1/4" holes in the head channel, using the two tapped holes in the quill support base to guide the drill. The cap must be removed to drill the holes, and it will be best to lay the machine on its back so the shims won't be dropped and mixed up. Be careful to keep all in order so you can reassemble as it was taken apart. Tap the holes in the head channel 5/16"-18, and enlarge the holes in the support base to 5/16". Install two 5/16"-18 X 1" bolts with lock washers and reassemble the cap before you drill the upper holes in the ears of the quill support base.

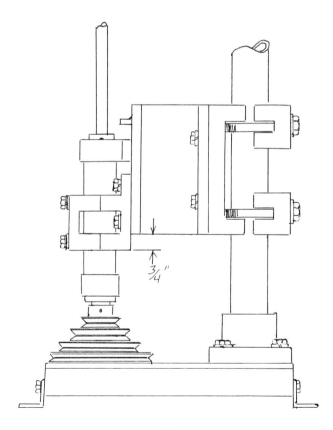

Check the alignment as you snug up the lower bolts, and be sure the cap and shims are properly installed as they were removed.

Step drill 1/4" holes through the ears and the head channel, tap the holes in the channel and enlarge the holes in the ears to 5/16". Install the remaining two bolts in the ears.

The cap must be removed again so that the feed stud can enter the hole in the upper quill bearing support. Reinstall the cap before you lock the set screw on the feed stud.

Now you can add or remove shims in the quill support until you have a smooth easy feed with just a tiny bit of play. The quill tube must be well oiled, and make notes as you add and remove shims so you don't get mixed up.

The quill guide will wear to a better fit after some use, and you can remove additional shims to bring it to a more perfect fit later.

It is possible that there may have been errors as the parts were machined, and your alignment may be something less than perfect. You can install shims between the head clamp base and the head channels at either the top or the bottom if the error seems significant.

It's more likely that all will be aligned to within a very small fraction of one degree, and that will meet the requirements of most home shop operations. It is bound to be a great improvement over free hand drilling in any case, and any parts that may need re-doing will be more easily done when the drill press is complete.

CHAPTER VI

THE SPINDLE DRIVE

The driven pulley is mounted on its own ball bearings and supported to run concentric with the spindle so that no belt tension is transferred to the spindle.
The keyway in the driven pulley is fit with a floating key to drive the spindle by a keyway cut in the upper end. The keyway can be cut on the miller, the shaper, or by hand with the simple tool described.

CUTTING THE KEYWAY BY HAND

This simple method is very old, and I've used it with success a number of times. It has been described in Popular Mechanics publications several times over the past 40 or 50 years. It's the easiest way to cut the keyway in the spindle if you have not built the miller of the shaper yet.
Five hacksaw blades are clamped in the simple wooden fixture, and it is used like a plane or a file to cut the keyway while the spindle is clamped to the bench top.
Either 18 tooth or 24 tooth blades will do the job, with 18 tooth being the better choice. Cheap carbon steel blades will work just fine for this job.
This tool with five blades cuts a keyway slightly less than 3/16" wide, and it takes only a bit of hand filing to widen it to 3/16" for a smooth sliding fit on the key.
The finished keyway is 3/16" wide, 3/32" deep and 5" long. There is no harm in going a bit deeper.

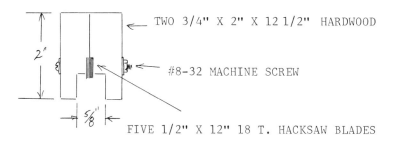

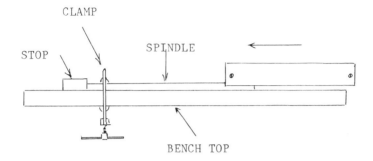

Use the tool as you would a file or hack saw. Lift a tiny bit on the back stroke, and make firm forward strokes as you hold the tool level with the bench. A single clamp and a stop block will prevent movement fo the shaft. It won't take long and you'll have a clean keyway.

MILLING THE KEYWAY

A simple cast aluminum holding fixture and a fly cutter will do the job. The holding fixture will find many applications on the miller, shaper, drill press, and also for bench work. Either the miller or the shaper will do the job with this fixture.

It's a simple pattern and a routine molding job, and it's mounted on the face plate with a length of shaft as a spacer to face off the bottom surface. The result is a V block with clamps that will mount round work parallel to its base.

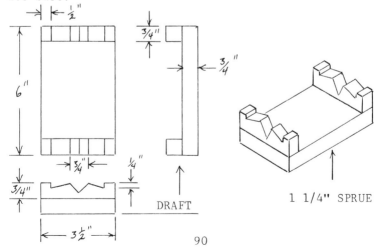

The V has a 90 degree included angle. Make the ends as near identical as possible, and mount them on the base of the pattern true square and parallel. Only minimum draft is needed, and a small fillet at inside corners.

Two 3/8" elongated holes are cut in the base of the casting for mounting on the work table, and two 5/16"-18 holes are tapped for mounting it on the face plate to machine the surface of the base.

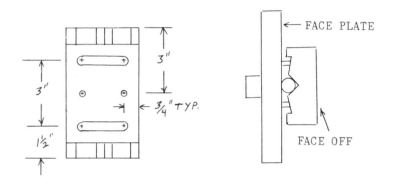

File the V blocks for good contact on the shaft, and mount it on the face plate with a 6" length of 5/8" shaft as a spacer and face off the base. Align the casting so it is near parallel to the face plate, and be careful not to over tighten the bolts.

Having been faced off with the shaft in the V block, the fixture will support the shaft parallel to the work table of the miller or the shaper.

The 3/8" elongated holes can be cut with a coping saw after drilling a 3/8" hole at each end of the slot.

Four 3/16" X 3/4" X 1" clamps are tightened with 1/4" cap screws to hold the work in the fixture.

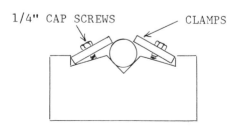

The spindle is clamped in the fixture, and carefully aligned parallel to the work table travel. The elongated holes in the base of the fixture allow for adjustment, and you can use the edge of the fly cutter to test on each end of the travel to align the work. You can use a small try square resting on the work table and against the side of the spindle to guide you as you position the cutter in the center of the spindle.

I made a simple stub holder for the millers 5/8" chuck, and fitted it with a 3/16" fly cutter to mill the keyway. This is the simplest form of cutter, and its general anatomy is like that of a boring tool. It requires back rake, end relief and side relief Wet hone the edge to razor sharpness. Such a cutter seems small, but a 3/16" wide cut in steel requires considerable power. I ran the mill at 270 RPM and made about 10 passes, each .010" deep. You won't gain any strength in the drive by cutting deeper than .100". Lubricate the tool with oil on each pass, and give it a slow steady feed.

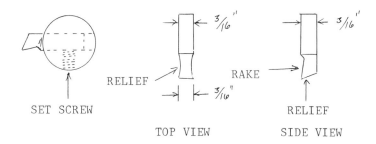

A shaper tool is ground to the same form, and the set up procedure is generally the same. Drill a shallow 3/16" hole at the end of the cut for tool clearance if you use the shaper to cut the keyway. Use the shapers lowest speed and feed about .005" on each stroke.

The keyway must be on the center line of the shaft, and it must be parallel to the center line. Take plenty of time to make a careful set up.

THE DRIVE KEY

This type of key is called a " feather " key. It fits closely in the pulley's keyway, and it has a lug that enters the set screw hole so it won't fall out.

Make it from a 1 1/2" length of 3/16" X 3/8" key stock. Cut it to approximate size with a hack saw, and finish with a file to fill the space of both keyways with just enough clearance for a smooth sliding fit between the pulley and the spindle. The spindle must move through the pulley for the full length of quill travel without binding.

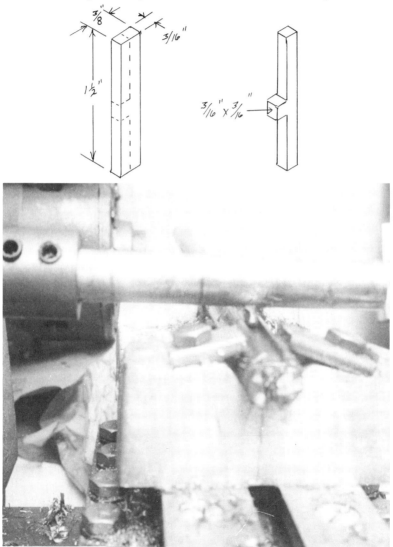

CUTTING THE KEYWAY WITH THE MILLER

FITTING THE CHUCK

If you purchase a chuck with a 5/8" straight bore, the spindle end need only have a flat ground or filed for the set screw to seat against. I used such a chuck from Gilliom on an earlier drill press, and it was entirely satisfactory.

The chuck you see in the photos was purchased from the Wholesale Tool Co. It came with a #1 morse taper adapter, which adapts it for use in the lathe tail stock, and its back is bored for a jacobs taper. It's an easy job to cut the jacobs taper on the drill press spindle.

You can mount the spindle in a pair of pillow blocks, drive it with the same pulley arrangement as for dressing the column, and form the taper with a file. You'll notice an extra bearing support in the photo of my set up for the column, and that was planned for cutting the taper. The job can also be done on the drill press after the motor is installed.

Just estimate the taper as compared visually with the tapered adapter, and file the rotating spindle to a near approximation. Stain the trial taper with prussian blue oil paint, hold it stationary, and rotate the chuck one turn by hand to rub the stain off the high spots. When you remove the chuck you will be able to see whether to increase or reduce the taper. When the taper is right, you will feel no side play and you'll have to rap the chuck with a hammer to get it off the spindle end.

I wanted a more perfectly concentric taper, so I removed the spindle form my lathe, installed a 3/4" X 5/8" bushing in the large end, and installed the drill press spindle to cut the jacobs taper with the compound. I made trial passes, testing as described above, until I had the compound angle set correctly, and then finished the taper with a few more light passes.

CAUTION:

THIS TYPE OF CHUCK MOUNT IS NOT SUITABLE FOR ANY PURPOSE BUT DRILLING. SUCH ACCESSORIES AS FLY CUTTERS, ROTARY PLANING HEADS, GRINDING WHEELS, SPIRAL END MILLS OR SUCH TOOLS THAT TEND TO DRAW OUT OF THE CHUCK CAN NOT BE USED SAFELY. A SUDDEN JAR CAN DISLODGE THE CHUCK AND SEND IT FLYING ABOUT YOUR SHOP. COMMERCIALLY BUILT MACHINES HAVE A LOCKING CHUCK MOUNT FOR SUCH ACCESSORIES.

CUTTING THE SPINDLE TAPER ON THE LATHE

THE THRUST COLLARS

The upper collar merely supports the weight of the spindle and the chuck, so it can be used with its single set screw against a small flat ground or filed on the spindle.

The lower collar opposes the force of drilling, so two additional set screws should be installed at 120 degree intervals. Form a shallow dimple in the spindle by drilling with a tap size drill through the set screw holes.

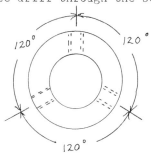

THE DRIVEN PULLEY

The belt drive applies considerable pressure on the pulleys, and the driven pulley is mounted on a hollow shaft that is supported independently of the spindle bearings so that the spindle won't be drawn out of alignment by belt tension.

The photo on page 73 shows the driven pulley mounted on the head, with its cast aluminum hub and the three castings that support it in alignment with the spindle.

I used a 5"-4"-3"-2" die cast pulley purchased from Sears tool catalog. Its 5/8" bore has a 3/16" keyway, and it is left intact. The back side of the pulley is hollow, and there are four reinforcement ribs which must be partly broken away. The pulley is then mounted on an arbor between centers and the recess of the first step is bored to make it concentric with the hub bore. These pulleys are of thin section, and only enough material is removed to produce a clean straight sided bore. Mine measured 3.800" at the start, and finished at 3.815", so the wall thickness was only reduced by .0075".

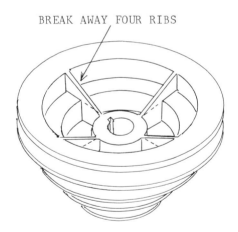
BREAK AWAY FOUR RIBS

THE PULLEY SPINDLE AND HUB

The hub is cast on a length of 3/4" black pipe which will become the pulleys hollow spindle. The pattern is a 4" disc of 3/4" pine, bored to an easy slip fit on the pipe. The protective coating is sanded off the pipe, it is packed with molding sand and rammed up in the mold as a core. The disc is drawn from the mold, but the pipe is left in place to become a part of the casting.

It will be best to mount the disc on the face plate with a couple of set screw collars as spacers, and bore the core hole accurately. Then you can fill the mounting holes with body putty to finish the pattern.

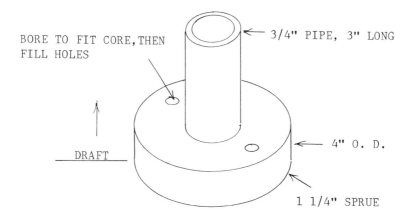

Pack the pipe core with sand, set it in its bore in the disc and ram up the drag. Roll over, set the sprue pin about 1" away at any point, and ram up the cope. Open the mold, rap and draw the disc, cut the gate and close up to pour. The pipe core will be a permanent part of the casting.

MACHINING THE HUB SPINDLE

Mount the casting in the V block angle plate by its pipe spindle, and face off the base of the hub. Drill two 1/4" holes in the hub and tap them 5/16"-18. Mount it on the face plate with a pair of set screw collars for spacers, just as you did for boring the core hole in the pattern.

Take plenty of time to center the pipe on the turning axis of the lathe. The pipe is only about .050" over 1" diameter, so there is little room for error here. Make sure the pipe is running parallel to the turning axis so it won't finish thin at any place. The exposed face of the hub is faced off to reduce its thickness to 5/8", and the diameter of the pipe is carefully reduced to 1". Cut the pipe off at 2" from the surface of the hub, and face off the end. Reduce the diameter of the hub to exactly the same diameter as the bored step of the pulley. Measure very carefully so that you get a snug fit. All of this work must be done without changing the set up so that all will be concentric.

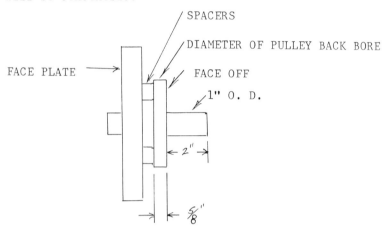

Finish the surface of the pipe spindle with a file until the 1" bored ball bearing will slip on easily without perceptible play.

Gently lock a 1" set screw collar on the spindle 1 3/4" from the hub to serve as a guide for the retaining ring groove. Rest the hack saw blade against the collar as you cut a shallow groove for the retainer.

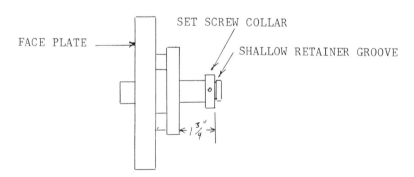

The depth of the retainer groove is only about .050", and the wall thickness of the pipe won't allow for very much more than that. Use the retainer as a test gauge as you make light cuts with the hack saw. You can face off the end of the spindle an additional amount when the groove is finished.

The hub should fit the back bore of the pulley without play, and a few light hammer blows should drive it against the shoulder of the bore. Two #8-32 flat head machine screws are countersunk in the groove of the pulley into tapped holes in the hub to fasten it permanently. The set screw collar is moved against the hub to serve as a spacer.

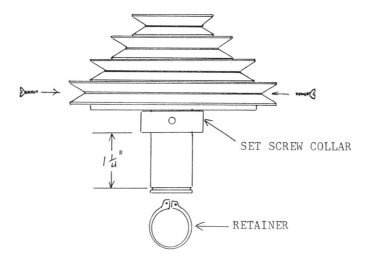

THE DRIVEN PULLEY BEARING SUPPORT

Similar to the column base casting. Only minimum draft is required on all outside surfaces, but the bore is steeply tapered to leave a clean core in the mold. Mount the pattern on the face plate with a couple of collars to bore the core hole, then fill the bolt holes with body putty to finish the pattern. It's a routine mold using a 1 1/4" sprue.

Machining is just like the column base casting. It's mounted on the face plate to face off the base, then it's inverted on the face plate to bore it to 2", leaving the small shoulder at the bottom of the bore.

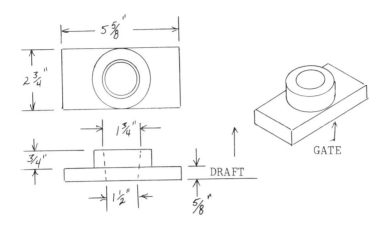

Tap two 5/16" holes in the base and mount the casting on the face plate to face off the base and enlarge the bore to 1 3/4" for a depth of about 1/4". Center the casting so the bore will be concentric with the outside diameter of the hub portion.

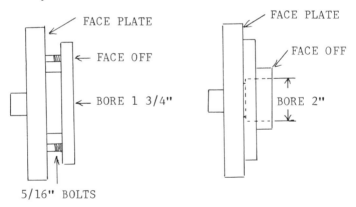

Invert the casting on the face plate, and again center it so the bore is concentric with the outside diameter of the hub portion. Face off the hub and bore it to 2" for a depth of 1 3/16" to leave a shoulder at the bottom of the bore.

The bearings are bored 1" X 2" O. D., and they are 1/2" thick. The shoulder at the bottom of the bore will retain them at the lower end, and a pair of 1/4" cap screws will retain them at the top.

It will be best to bore to a slip fit rather than a force fit.

A spacing ring is made from 1/16" X 3/16" steel to separate the bearings.

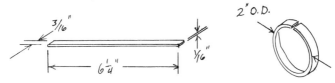

One bearing is pushed into the bore, then the spacing ring, and the second bearing is pushed in and fastened with the two 1/4" cap screws with flat washers and lock washers. Push the hollow spindle of the pulley through the bearings and install the retaining ring. Adjust the set screw collar to take up the end play and lock the set screw.

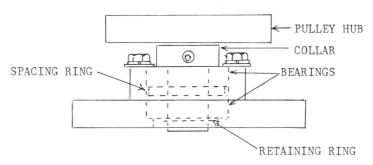

THE HEAD SIDE CASTINGS

These provide mounting for the pulley bearing support and the switch housing. There is a right and left hand member.

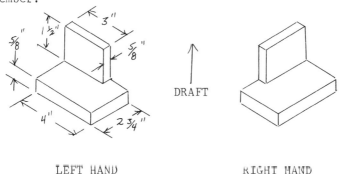

The dimensions of both patterns are the same. 5/8" plywood will work OK for these simple patterns. Minimum draft is sufficient, and wipe a small fillet on the inside junction. It's a routine mold fed with a 1 1/4" sprue at any convenient point.

MACHINING THE HEAD SIDE CASTINGS

Tap a 5/16" hole in the center of the largest surface of each casting. Mount the casting on the face plate with a stack of set screw collars for spacers to clear the leg, and face off the surface smooth. The faced off surface is then mounted on the angle plate so that the leg can be faced off at right angles to it.

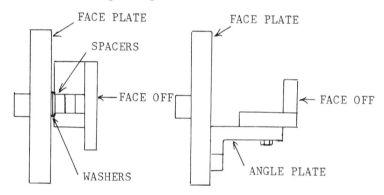

Finally, rotate the casting 1/4 turn, and align the faced off leg at exact right angles to the face plate to face off the top surface. The bottom surface of the leg is notched to clear the mounting ears of the quill support, and it is left as cast.

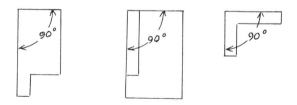

INSTALLING THE DRIVEN PULLEY

   Having been machined true square on three surfaces, the head side castings will support the driven pulley bearing support at right angles to the front vertical surface of the head channels. Refer to the photo on page 73 for a clear view of the installation.
   Make a trial assembly by clamping both side castings in their approximate location on the head channels with 4" C clamps. Install the key in the driven pulley and slip the entire pulley assembly and bearing support onto the spindle to rest on the side castings. Tap the side castings with light hammer blows as you adjust their position until the spindle will move freely through the pulley as the quill is fed. Make sure there is clearance between the pulley spindle and the upper set screw collar of the quill assembly. When all is well aligned, the bearing support will rest squarely on the side castings, and the spindle will slip through the pulley without perceptible drag as the quill is fed. Take plenty of time to do all of this work carefully.
   If there has been an error in machining any of the head parts it is likely to show up now. You can re-machine the defective part, or you can install shims to compensate. An error of 1 degree or less is not likely to be apparent.
   The side castings are each held to the channels with two 1/4"-20 X 1" cap screws with lock washers. As in all such assemblies, drill the tap hole through both members, tap the hole in the channel, and enlarge the hole in the side casting to 1/4". Install one bolt completely before you drill the second hole, and install the second bolt before you drill the third hole. Carefully check alignment before drilling each hole.
   Measure the distance between the castings both top and bottom to make sure they are parallel before you install the third and fourth bolts.
   You'll have to use a 6" long bit to drill the 3/16" tap holes because of interference from the quill and the pulley.
   When the side channels are permanently bolted in place a 1/4"-20 X 1" cap screw with lock washer is installed in each corner of the driven pulley bearing support to fasten it to the side castings.

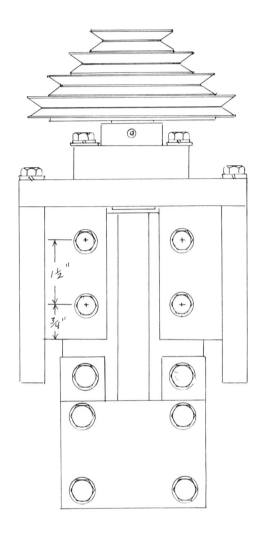

FRONT VIEW WITH QUILL REMOVED

   The view above is as though all were cleared away so you can see the relative position of the side castings, the pulley bearing support casting and the quill guide.
   It may be neccessary to remove the pulley to install the bolts through the bearing support into the side castings, but I was just barely able to install mine with the pulley in place. If you must remove the pulley, be sure

the bearing support casting is well clamped to the side castings in proper alignment before you remove it. Check alignment by installing the pulley before you drill the second bolt hole.

THE SWITCH HOUSING

A simple box with sloping sides, and it has ears to bolt to the front surface of the side castings. The opening is the right size for a standard wall switch.

All four sides slope from the base to the smaller dimension, and it is of 1/4" section thickness throughout, so it has draft both inside and out. You could make it of 1/4" plywood, but pine stock is easier to work. Fasten the parts together with glue and brads, and sand the top and bottom surfaces to bring them flat and even. Fill any flaws with body putty and sand smooth all over both inside and out.

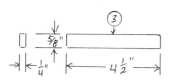

SIDES, MAKE 2

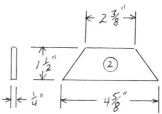

ENDS, MAKE 2

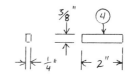

EARS, MAKE 2

LUGS, MAKE 2

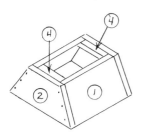

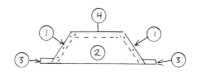

Cut out all of the parts to the size and shape in the sketches. Fasten the ends to the sides with glue and small brads, leaving an equal amount of the sides to protrude at both the top and bottom. Use coarse sand paper over a block of wood to sand the top and bottom surfaces flat and parallel. Cut the ends of the lugs to the proper angle to fit them to the inside of the box so they will be even with the top surface, and install them with glue and brads. Form the proper angle on the edges of the ears and fasten them to the sides to complete the pattern. Wipe a small fillet at all inside corners, fill any flaws, and make sure the lugs and ears have a slight amount of draft. Finally, make sure the pattern will rest squarely on the molding board and seal with varnish or lacquer.

This must be a double roll mold so that the large green sand core will rest on the parting face of the drag rather than hang suspended from the cope sand. Cut a 2 1/8" X 4" plate from thin sheet metal to cover the top opening. Lay the pattern on the molding board and set the cover on it. Set a 1 1/4" sprue pin about 1 1/2" away from either end of the pattern and ram up the cope. Vent the cope with the wire, finish the sprue opening and pull out the sprue pin. Rub in a bottom board and roll over the cope. Ram up the drag, being sure to fill the hollow portion of the pattern, but don't ram it any harder than the main body. Vent the drag generously, especially through the core area, rub in a bottom board and roll over the entire mold. Open the mold carefully, and set the cope on edge behind the drag.

It is most likely that the pattern will remain on the drag, in which case simply rap it a bit and lift it from the green sand core. If it remains in the cope, lay the cope down on a bottom board to swab, rap and draw it. Cut the gate in the cope rather than in the drag. Be sure to remove the sheet metal cover, clean up the mold, and close up to pour.

The housing is fastened to the front surface of the head side castings with four 1/4" screws or bolts tapped into the side castings.

Break the plaster ears from a standard 20 amp wall switch, and tap two #6-32 holes in the lugs to mount the switch. A stamped steel switch plate will finish the job.

CHAPTER VII

THE MOTOR DRIVE

At this point you can install the motor drive with a 4"-3"-2" pulley on the motor and you will have a machine that is adequate for many home shop purposes. It will be much better if you add the second stage speed reduction, for it is almost certain that you will want to do some work that requires the lower speed range. The single stage reduction gives a low speed of 690 RPM, and that is adequate for drilling a 3/8" hole in steel. The second stage will reduce the low speed to about 275 RPM so that you can drill up to 3/4" holes in steel without burning up the bit.

In either case the motor mounting bracket and all other parts of the motor drive will be the same.

THE MOTOR MOUNTING BRACKET

A rectangle of 1/4" plywood with a 3/4" X 1" X 6" pine block glued and bradded to one end forms the pattern.

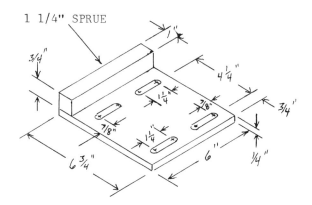

Only minimum draft is needed at the outside edges, but the elongated holes need more draft. They are 5/16" wide at the parting line and 7/16" wide at the top surface. It is difficult to form such tapered holes, especially in a material like fir plywood. You can make a tapered wooden form, cut the elongated hole over size, and mold it to shape with body putty. Make the form very smooth and coat it with parafin so the putty won't bond to it. Just one form will serve to shape all the holes, as the body putty sets up in just a few minutes.

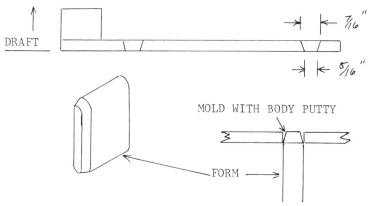

The holes must be finished smooth so that the small cores won't be broken away when the pattern is drawn.

Molding the bracket is routine, but make sure that you press the elongated holes full of sand to leave firm cores in the mold. You can press a couple of small brads into each core before you ram up the cope to reinforce them.

INSTALL THE MOUNTING BRACKET

File the elongated holes to a smooth fit on the 5/16" carriage bolts that will mount the motor.

Fasten a 3 1/2" butt hinge to the rail with 1/4" cap screws with lock washers.

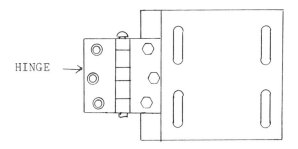

HINGE

The other half of the hinge is fastened to the left side of the head clamp blocks with two 1/4" cap screws with lock washers.

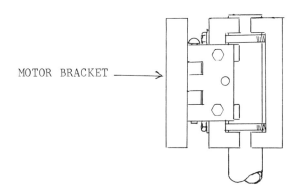

MOTOR BRACKET

THE BELT TENSION LEVER

A single motion of the lever moves the motor forward to release the belt tension for fast speed changes. Return the lever and the belt is tightened.

The mechanism must pivot on both vertical and horizontal planes, so the lever and link are mounted on universal brackets.

The bracket parts are made of scraps of 1/8" X 1 1/4" angle iron, and the lever and link are 1/8" X 1" strap. The pivots are 1/4" iron rivets, and the lever stop is a 1/4" fillister head screw with a lock washer. Make all of the parts as detailed. Only one of each is required.

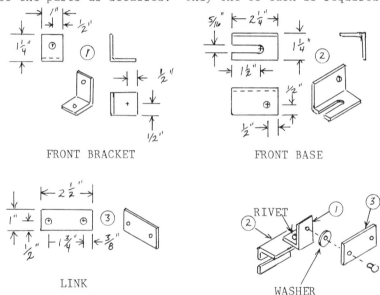

FRONT BRACKET

FRONT BASE

LINK

RIVET

WASHER

Assemble the front bracket to the front base with a 1/4" rivet. Join the link to the front bracket with a 1/4" rivet and a small flat washer between the joint. A standard 3/16" washer can be reamed to a close fit. There is no washer in the bracket to base joint.

The elongated hole in the base is easily formed by drilling a 5/16" hole and cutting away the waste with a hack saw.

Peen the rivets well to make a tight joint. Don't oil the joints at this time.

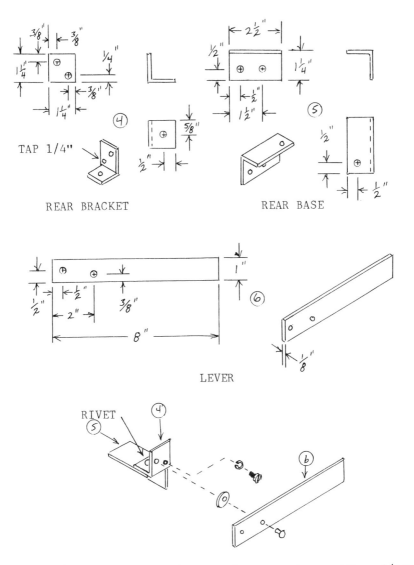

REAR BRACKET

REAR BASE

LEVER

Join the rear bracket to the rear base with a 1/4" rivet. Join the lever to the rear bracket with a 1/4" rivet and a small flat washer between the joint. Install a 1/4" fillister head screw in the tapped hole, which will be the lever stop.

Join the link to the lever with a 1/4" rivet with a washer in the joint and the assembly is complete. Wash it with solvent and give it a coat or two of aluminum paint for good appearance. When the paint is dry, put a drop of oil on each joint and work them around to free them up.

Drill two 1/4" holes through the motor mounting bracket to match the holes in the rear base bracket. Countersink them from the rear and install the rear base with flat head screws, nuts and lock washers.

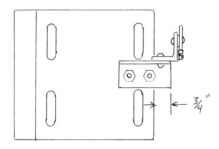

The front base bracket is fastened to the head clamp base with a 5/16" cap screw, flat washer and lock washer. Position the base on the head clamp so that the pivot of the link and lever is about 1/8" above a line through the lever pivot and the front link pivot when the lever rests against the stop. Mark the center of the elongated hole and drill and tap a 5/16"-18 hole in the head clamp.

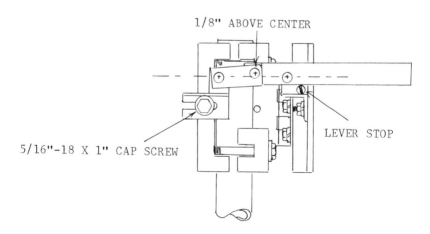

The elongated hole in the front base permits adjustment for belt length variations. When the lever is raised the motor is drawn forward to loosen the belt for speed changes. When the lever is returned to the stop the motor is drawn back to tighten the belt. When the lever link pivot passes over center, the linkage is locked by the belt tension.

THE BELT GUARD

There is plenty of good argument for installing a full belt guard, though I notice that they are often removed from machines that come equipped with them. Install the partial guard at the front at the very least. You may be able to remember to keep your fingers out of the drive, but there is a serious danger of getting your hair caught. You also need protection in case a belt breaks or frays.

The angle iron brackets support both the belt guard and the countershaft bracket. They are 1/8" X 1 1/4", and they are fastened through the slotted holes by the same bolts that fasten the driven pulley bearing support. Locate the slots so that the front of the angles are in line with the vertical surface of the switch housing. The rear angle brackets are bolted to the sides of the head channels. See the photo on page 107.

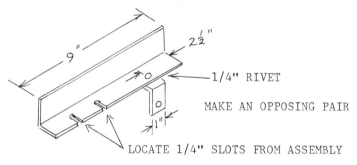

1/4" RIVET

MAKE AN OPPOSING PAIR

LOCATE 1/4" SLOTS FROM ASSEMBLY

The 9" long angle bracket is for the front guard only. If you elect the full belt guard, make the brackets 20" long with the slots and the side brackets in the same relative position.

Loosen the pulley bearing support bolts and slip the brackets in place for a trial assemble to locate the rear side brackets. You may have to use washers between the brackets and the channels to bring them parallel. Space the brackets 6" apart and parallel front to back.

I painted my angle brackets with aluminum paint, and made the guard of heavy embossed aluminum such as is used for the solid panels in aluminum storm doors.

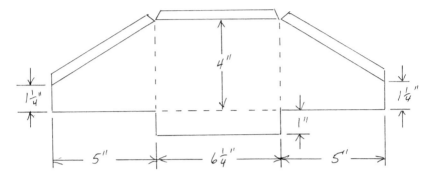

Fold the hem to kill the sharp edge, and bend up at the dotted lines. Fasten the guard to the angle brackets with four 1/4" bolts with lock washers.

If you have elected to install the full guard, make a box of the same material with the rear end open and the front end cut to match the angle of the front guard.

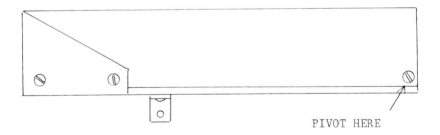

PIVOT HERE

A hole will have to be cut in the top of the full guard for the spindle to pass through.

THE COUNTERSHAFT

The bracket casting for the countershaft is so much like the pulley bearing support that little need be said about casting it and machining it. The pattern is simple, and molding is routine. Feed with a 1 1/4" sprue.

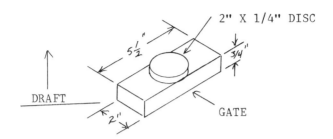

You could bore the pattern for a 1" core, but it is hardly worth the labor for a one only job. It is about as easy to drill a starting hole to bore the casting.

Tap 5/16"-18 holes in diagonally opposite corners of the casting and bolt it to the face plate with set screw collars for spacers. Bore the center to 1.125" all the way through, and counterbore to 1.375" to a depth of 7/8".

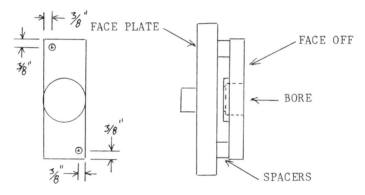

You can bore about one half thousandth undersize for a force fit on the bearings, or you can retain them with screws and washers as on the pulley bearings.

The bearings are 5/8" bored ball bearings just as the quill bearings. The O. D. is 1.375", and they are 5/16" thick. Make a spacing ring from 1/16" X 1/4" steel to separate the bearings.

The spindle is a 3 1/2" length of 5/8" cold rolled steel with hack saw grooves for the retaining rings. I mounted mine in the spindle chuck of the miller to cut the grooves. You can do the same job by mounting a length of shaft in a pair of pillow blocks to cut the grooves, then cut it to length. The grooves are spaced 7/8", and you can use a set screw collar to guide the hack saw, just as you did for the driven pulley spindle. Grind a flat on the shaft for the pulley set screw to seat against.

SPACING RING

SPINDLE   COUNTERSHAFT ASSEMBLY

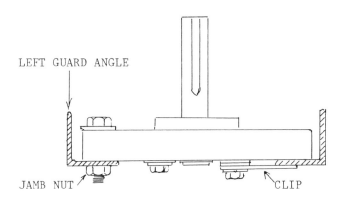

CUTAWAY VIEW AT FRONT

The enlarged drawing on page 116 shows the assembly as though it were viewed from the front. Enlarge one of the holes in the countershaft bracket to 5/16". Make a clip of 1/8" X 3/4" steel and bolt it to the bottom of the bracket with washers to space it for a smooth sliding fit on the right hand guard angle. You can use one of the bearing retaining bolts, or tap a separate hole in the bracket. Tap a 5/16" hole in the left hand guard angle, at about 1 1/2" from the rear. Install the bracket to the angle with a 5/16"-18 X 1 1/4" cap screw and flat washer. Adjust to a good fit and lock the bolt with a jamb nut.

Now you can install the second step pulley and the motor with its pulley. My machine used a 23" belt and a 27" belt, but you had best measure to find your belt size.

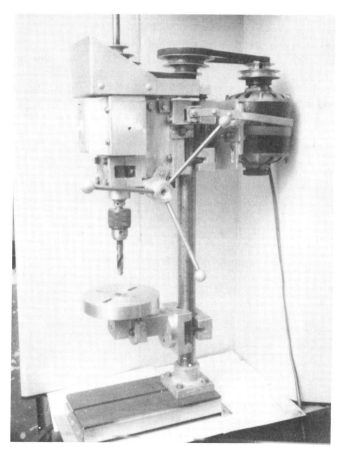

CHAPTER VIII

THE WORK TABLE

The original work table support arm, seen in the photo on page 31, was a single casting, which was to be machined between centers on the lathe and bored on the miller. I suddenly realized that many of you may not have built the miller, so I separated the arm into two castings so that the entire job can be done on the lathe.

THE TABLE CASTING

It looks just like a lathe face plate, and it is machined just like a face plate. It will be best to turn the pattern to size on the lathe so you will be sure the casting will be within the lathes capacity.

Make the face and rim of 5/8" plywood, and the hub of 2" discs of 3/4" pine. Make a wooden form for the elongated holes, just as you did for the motor mounting bracket, and mold them with body putty. They are 3/8" wide at the parting plane and 1/2" wide at the top. Assemble the parts with glue and brads, and mount the pattern on the lathe face plate to machine the rim and hub concentric.

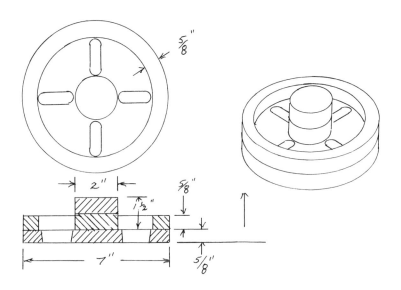

Molding is routine, but don't forget to press sand into the elongated holes as you ram up the drag. It helps to push a couple of brads into each of the small cores before you ram up the cope. This casting requires a heavy riser in the center to feed the hub. It will be easiest to set a 1 1/2" sprue at the center, which will serve as the pouring gate and the riser. This is called a " pop " gate.

MACHINING THE WORK TABLE

The sprue will be on the face of the casting, and an 18 tooth hack saw blade will make short work of cutting it off. Mount the casting on the lathe face plate with four bolts through the elongated holes, using set screw collars for spacers so you can bore through without touching the lathe face plate. Center it carefully and bore it through to exactly 1". Install a 5/16" or 3/8" set screw in the hub and mount it on the 1" mandrel to face it off and machine the rim.

The shank that mounts the table on the arm is a 3 3/4" length of 1" cold rolled steel. You can mount the finished work table on the face plate to serve as a set screw chuck to hole the shank as you face off the ends. Grind a flat along the length of the shank for the set screws to seat against.

THE WORK TABLE SUPPORT ARM PATTERN

A bit unusual in shape, but it's designed especially for machining on the small lathe. It's a split pattern with the drag half in two pieces. Draft is provided by the sloping vertical members and the added wedges.

Make the center body of 1/2" pine stock. Note that it tapers 1/8" on both sides and the ends for draft both inside and outside.

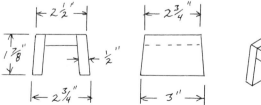

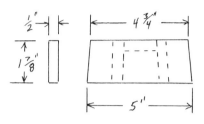 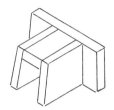

Add the front end, which forms the ears for mounting the table shank bracket as shown above, and add a wedge to the back side of each ear for draft as shown below.

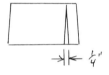 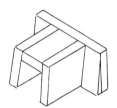

The back end is a half circle of 3/4" pine stock, and the drag half is a half circle of the same radius. Note that the drag half is tapered for draft, while the cope half is not. A wedge is added to each wing of the circular part of the cope pattern for draft.

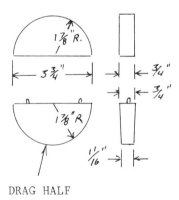 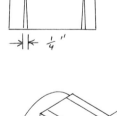

DRAG HALF

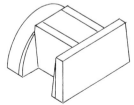

A second half circular part is added to the front end to provide support while machining. It is the second part of the drag half, and it's tapered for draft. Both circular parts are aligned with the cope half by small pins.

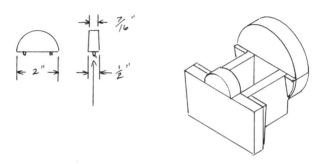

The pins can be made of small brads, and a brad of the same size can be used to drill the mating holes in the cope half of the pattern. They need extend mo more than 1/8", and they must fit the mating holes freely so that the drag half patterns will separate easily and remain in the drag when the mold is opened.

Wipe a small fillet on all inside corners, fill any flaws, and sand smooth both inside and outside before you seal with lacquer or varnish. Make sure the drag half separates easily after sealing.

MOLDING THE SORK TABLE SUPPORT ARM

This is another double roll mold because we want the green sand core to rest on the drag rather than hang from the cope sand.

Lay the pattern on the molding board and set a 1 1/4" sprue pin about 1 1/2" away from the circular end. Ram up the cope, using the drag half of the flask so the pegs won't interfere. Be sure to fill all the corners, and ram uniformly. Vent generously with the wire. Finish the sprue opening to a funnel shape, and remove the sprue pin. Rub in a bottom board and roll over the cope. This is the first roll.

Set the drag half patterns in place and ram up the drag. Be sure to press sand into the core area, but don't ram it any harder than the main body of sand. Vent the core area generously with the wire. Rub in a bottom board and roll over the entire mold. This is the second roll.

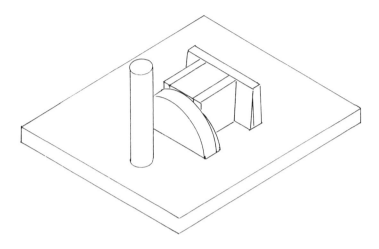

Push a 1/4" rod through the cope sand and rap the pattern before you open the mold. You can push sand back in the hole left by the rod, or you can leave it open for a vent. Lift the cope straight up and set it on edge behind the drag. Rap the pattern all around and lift it off the core. Swab the drag pattern parts and grasp them by their pins to draw them from the mold. Clean out the sprue and cut a full sized gate in both the cope and the drag. Use a light to look into the cavities, and lift out any loose sand with a wet swab. Clean up the entire mold carefully before closing up to pour.

MACHINING THE WORK TABLE SUPPORT ARM

The gate will be easy to cut off with an 18 tooth hack saw blade. The first step is to prepare 60 degree centers to mount the casting between centers on the lathe.

The object here is to locate the center of the large circular portion and transfer it to the opposite end so the work will be mounted on its true axis. Some error is certain, for we lack a machined surface as a reference, and precision layout equipment is not likely to be available. The top surface is the best reference we have, so scribe a line along its center, parallel to the sides.

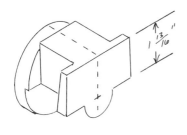 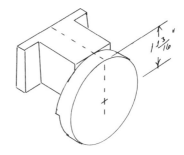

Drop the line on each end at 90 degrees to the top of the casting. Measure 1 13/16" from the top surface to the center of each end. Center drill each end and mount the casting between centers to face off the circular end.

A 5/16" cap screw with nuts on both front and back of the face plate will serve as a dog to drive the work. The plastic electrical tape seen in the photo is to hold the ear of the casting against the dog.

When the circular end is faced off, tap two 5/16" holes on the vertical center line and bolt the casting to the face plate to face off the opposite end. Space the holes about 1 1/4" above and below center.

When the front end is faced off, scribe lines from the center at 45 degrees from the top surface, and tap two 5/16" holes 3/8" from the top.

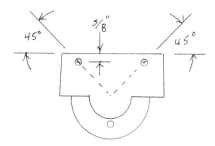

Use these holes to bolt the casting to the face plate, and add a clamp of 1/4" X 1" steel as seen in the photo.

Bring up the tail center to center the casting on the face plate, drill a 5/16" starting hole, and bore to a close fit on a 3/8" cap screw. Don't change the set up until the circular portion has been reduced to size for the protractor scale.

THE PROTRACTOR SCALE

If you have built the miller or the shaper the steel rule used for the protractor graduations is familiar. As there are 360 degrees in a full circle, we only need apply 360 equal divisions to the circular base of the arm and we have increments of one degree each. A flexible 6" steel rule has divisions of 1/32", and that's a convenient readable scale. Divide 360 by 32, and the answer is 11.25. A circle with a circumfrence of 11 1/4" will measure 1/32" for each one degree division. Divide 11.25 by 3.1416 to find the diameter, and you have approximately 3.581. The thickness of the rule is about .020", so the diameter is reduced an additional .040" so that the finished size including the thickness of the rule is 3.581". Then each 1/32" division on the rule will represent a one degree division on the circle. There are only 192 1/32" divisions on the 6" rule, but you only need 180, so it serves well for the need at hand. Subtract twice the thickness of your rule from 3.581", and that will be the finished size of the circular part of the work table support arm. The size is the same for a rule with 1/16" divisions, but of course the divisions will each represent two degrees. A 1/8" scale would give four degree divisions. Fasten the rule to the top half of the circle with two small machine screws, and fasten an adjustable witness mark plate to the top of the clamp base. The support arm is fastened to the clamp base with a 3/8"-16 cap screw with a flat washer.

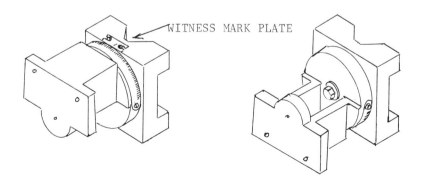
WITNESS MARK PLATE

It's a simple matter to loosen the cap screw and rotate the arm through any angle from vertical to horizontal. The scale divisions are likely to be something less than perfectly accurate, but they will be within a fraction of one degree, and that is close enough for most home shop work.

THE TABLE SHANK BRACKET CASTING

This is the final step in construction, and it's an easy casting and machining job. The pattern requires minimum draft, and it's fed with a 1 1/4" sprue at either side and a 1 1/4" riser at the opposite side.

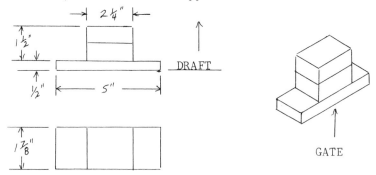

MACHINING THE TABLE SHANK BRACKET

Drill a 1/4" hole through the center of the casting, and tap 5/16" threads from both top and bottom. Bolt the casting to the face plate to face off the smaller surface, then invert it to face off the larger surface.

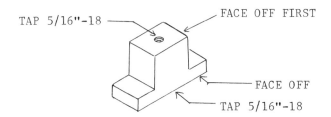

Clamp the shank bracket to the support arm and install a 5/16" cap screw in the center of each ear. Install the assembly on the column and chuck a center drill bit in the chuck to drill a center in the bracket.

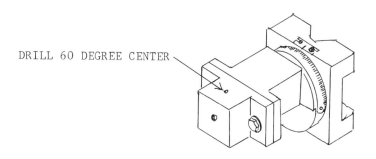

DRILL 60 DEGREE CENTER

Mount the bracket to the face plate with the angle plate and bring up the tail center to position it for the boring operation. Drill a starting hole with the tail stock chuck and bore the bracket to 1" for a close fit on the work table shank.

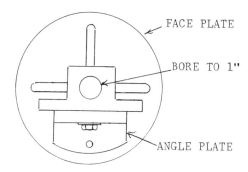

FACE PLATE

BORE TO 1"

ANGLE PLATE

Install the bored bracket on the support arm and install the work table shank in the bracket, locking it with a set screw in the front hole in the bracket.

Congratulations, you have converted your scrap box to a practical drill press. It, together with your lathe and the miller or the shaper will make easy work of building the accessories in book #6. You have only begun to enjoy the freedom and creative potential of your own machine shop. There is nothing to stop you now.

INDEX TO MAJOR SUBHEADINGS

BELT GUARD	114
BELT TENSION LEVER	111
BORING TOOLS	24
CABLE WINCH FEED	53
CHAIN DRIVE QUILL FEED	52
CHUCK FITTING	95
COLUMN BASE CASTING	35
COUNTERSHAFT	115
CROSS SHAFT	63
CROSS SHAFT BEARING SUPPORTS	56
DRAFT	8
DRESSING THE COLUMN	32
DRIVE KEY	93
DRIVEN PULLEY	97
FILLETS	9
HAND WHEEL	68
HEAD CHANNEL CASTINGS	47
HEAD CLAMP	44
HEAD SIDE CASTINGS	102
IDLER PULLEY	65
MOLDING	10
MOTOR MOUNT BRACKET	109
MOUNTING BASE	40
PARTING PLANE	9
PATTERN MAKING	8
PLAIN ANGLE PLATE	13
PROTRACTOR SCALE	126
PULLEY SPINDLE AND HUB	97
QUILL BEARINGS	74
QUILL GUIDE	82
QUILL TUBE	80
RETURN SPRING BRACKET	67
RISERS	11
SPINDLE KEYWAY	90
SPRUE	11
SWITCH HOUSING	106
TABLE CASTING	119
TABLE SHANK BRACKET	127
THRUST COLLARS	96
V BLOCK ANGLE PLATE	19
WORK TABLE SUPPORT ARM	120
WORK TABLE CLAMP	44